Developing Computational Fluency

James Burnett

MATHEMENTALS
Developing Computational Fluency

Grade 5

Author: James Burnett
Project editor: Beth Lewis
Illustrations and design: Brett Cox
Cover design: Brett Cox

For more information, contact:

USA
T 1-888-674-4601
F 1-888-674-4604
E info@origomath.com
W www.origomath.com

Canada
E info@origoeducation.com
W www.origoeducation.com

ISBN: 978 1 876842 32 1

10 9 8 7 6 5 4 3

CONTENTS

INTRODUCTION

Computational Fluency

The elementary school mathematics curriculum has long focused on the development of formal algorithms to add, subtract, multiply, and divide whole numbers, fractions, and decimals. However, todays students require more than a basic memorization of steps and rules for computation. This is the view of the National Council of Teachers of Mathematics (1) which recently emphasized 'computational fluency' as an essential goal for all students.

The Principles and Standards for School Mathematics described computational fluency as a "connection between conceptual understanding and computational proficiency". Conceptual understanding of computation is knowledge of place value, number relationships, and properties of operations. Computational proficiency is characterized by three main ideas: **efficiency**, **accuracy**, and **flexibility**.

A fluent student will adopt an **efficient** strategy that can be carried out easily without getting bogged down in a series of complex steps. A fluent student will have **accurate** recall of number facts and of important number combinations and show concern for checking results. Computationally fluent students are **flexible** thinkers. They can draw from a range of methods such as mental computation, estimation, calculators, and paper-and-pencil algorithms to solve the problem at hand.

Many resources are available to assist teachers in developing student's skills in estimation, formal written algorithms and in the use of the calculator. There is however a lack of quality materials that assist teachers in developing a range of mental computation strategies. The activities in ***Mathementals*** are designed to achieve this aim.

The Mathementals Series

Mathementals is a set of six reproducible blackline master books designed to help students develop a range of mental computation strategies. Each book provides appropriate learning experiences with numbers and operations for students working at that level.

The books are divided into sections that focus on one operation. Each section has several ***Warm Up*** and ***Work Out*** exercises to provide the guided instruction and written practice. These double-page spreads develop one specific idea within a mental strategy. A range of number models, such as number lines, ten-frames, base-10 blocks, money, and hundred charts, are used throughout these exercises to imitate the thinking required for certain strategies and to give students practice in demonstrating this thinking. ***Check Ups*** are provided at various points throughout each book. Each of these is followed by a puzzle or game to engage students in mental exercises 'just for fun'. Answers are given at the end of each section.

How to Use this Book

The activities within each book have been developmentally sequenced. This allows teachers to work through the book from the front to the back cover. Alternatively, a confident teacher may want to develop his or her own sequence by drawing on related activities from the different sections of the book. For example, after completing the doubling activities for addition, the teacher may want to focus on the idea of using related doubles to subtract.

The instructional ***Warm Up*** page can be reproduced as an overhead transparency or as a class handout. The students should be 'walked' through the activity on this page, and encouraged to share their strategies and methods of computation.

The ***Mathementals***' character frequently appears with clear, open questions to prompt whole class discussion. There is evidence to suggest that peers can help each other progress from simple to more efficient strategies (2), so teachers should allow plenty of time for students to share their thinking with the class. As there is no one correct strategy, students should be encouraged to use the strategy that works best for them.

After discussing the ***Warm Up*** page, the students can 'sharpen' their mental skills by independently completing the ***Work Out*** page. This can be done in school time or as set homework. It is important to correct this page as a class. This will give the students further opportunities to share their strategies and to explain their thinking.

Assessment

It is important to know how confident and competent students are at using a particular strategy. This information can be used to judge whether they need further experiences with that strategy, or if they are ready to progress to the next strategy.

Rubrics are particularly helpful in assessing students' mathematical proficiency in open tasks, such as those that are often used to develop skills in mental computation. The rubric below offers a guide for tracking student progress through the ***Warm Up***, ***Work Out***, and ***Check Up*** pages. The findings can be recorded on the student progress summary provided on page 6.

A	The student mentally calculates all examples accurately. The student uses efficient strategies and is able to fully explain his/her thinking and reasoning. The student can describe more than one strategy to solve problems that are alike.
B	The student mentally calculates most examples accurately. The student generally uses efficient strategies and is able to explain his/her thinking and reasoning. The student can sometimes describe more than one strategy to solve problems that are alike.
C	The student mentally calculates some examples accurately. The student uses relatively inefficient strategies and has limited ability to explain his/her thinking and reasoning. The student cannot describe more than one strategy to solve problems that are alike.
D	The student mentally calculates most/all examples inaccurately. The student uses inefficient strategies and has poor/no ability to explain his/her thinking and reasoning. The student cannot describe more than one strategy to solve problems that are alike.

References

1. National Council of Teachers of Mathematics. (2000). *Principles and standards for school mathematics.* Reston, VA: Author.
2. Noddings, N. (1985). Small groups as a setting for research on mathematical problem solving. In E. Silver (Ed.), *Teaching and learning mathematical problem solving: Multiple research perspectives.* Hillsdale, NJ: Lawrence Erlbaum Associates.

Student Progress Summary

Name: ______________________

Refer to the rubric shown on page 5.
Place a ✓ or write the date in the appropriate column.

Strategy	#	Lesson	A	B	C	D
ADDITION Strategies	1	Count on – *regrouping*				
	2	Use doubles – *tenths with no regrouping*				
	3	Add the parts – *tenths and hundredths with no regrouping*				
	4	Use place value – *dollars and cents with no regrouping*				
	5	Round or adjust – *tenths*				
	6	Round or adjust – *dollars and cents*				
	7	Make a whole – *fractions*				
	8	Use compatible pairs – *four addends involving tenths*				
		Check Up 1				
SUBTRACTION Strategies	9	Count back – *from any number with regrouping*				
	10	Count on – *to a multiple of 10*				
	11	Count on – *dollars and cents*				
	12	Subtract the parts – *dollars and cents*				
	13	Use place value – *tenths with no regrouping*				
	14	Round or adjust – *rounding the subtrahend*				
	15	Round or adjust – *dollars and cents*				
	16	Exploring multiple methods – *with fractions*				
		Check Up 2				
MULTIPLICATION Strategies	17	Use doubles – *multiplying decimal fractions by four*				
	18	Double and halve				
	19	Double and halve				
	20	Use place value – *multiplying two two-digit numbers*				
	21	Use compatible pairs – *four factors*				
	22	Use factors – *of two two-digit numbers*				
	23	Use division – *halves and fourths*				
	24	Use division – *thirds and sixths*				
	25	Use division – *fifths and tenths*				
		Check Up 3				
DIVISION Strategies	26	Halve – *three-digit numbers*				
	27	Divide the parts – *three-digit numbers*				
	28	Divide the parts – *dollars and cents*				
	29	Break up the dividend – *three-digit numbers*				
	30	Round or adjust – *to divide by 5, 25, and 50*				
		Check Up 4				

ADDITION STRATEGIES

Count on

167 + 26 is the same as 167 + 10 + 10 + 6

Use doubles

6.4 + 6.4 is the same as double 6 + double 0.4

Add the parts

$32 + $6.23 is the same as $32 + $6 + 23¢

Use place value

$23.60 + $4.35 is the same as ($23 + $4) + (60¢ + 35¢)

Round or adjust

7.8 + 3.6 is the same as 8 + 3.4

$11.95 + $13.95 is the same as $12 + $13.90 or $12 + $14 – 10¢

Make a whole

$\frac{1}{2} + \frac{3}{4}$ *is the same as* $\frac{2}{4} + \frac{3}{4}$

Use compatible pairs

1.9 + 2.7 + 3.3 + 4.1 is the same as (1.9 + 4.1) + (2.7 + 3.3)

WaRM UP 1

Name: ______________________

The librarian bought 167 fiction books and 26 non-fiction books. How many new books in all?

1. **a.** Draw jumps on this number line to show how you could count on to figure out the total.

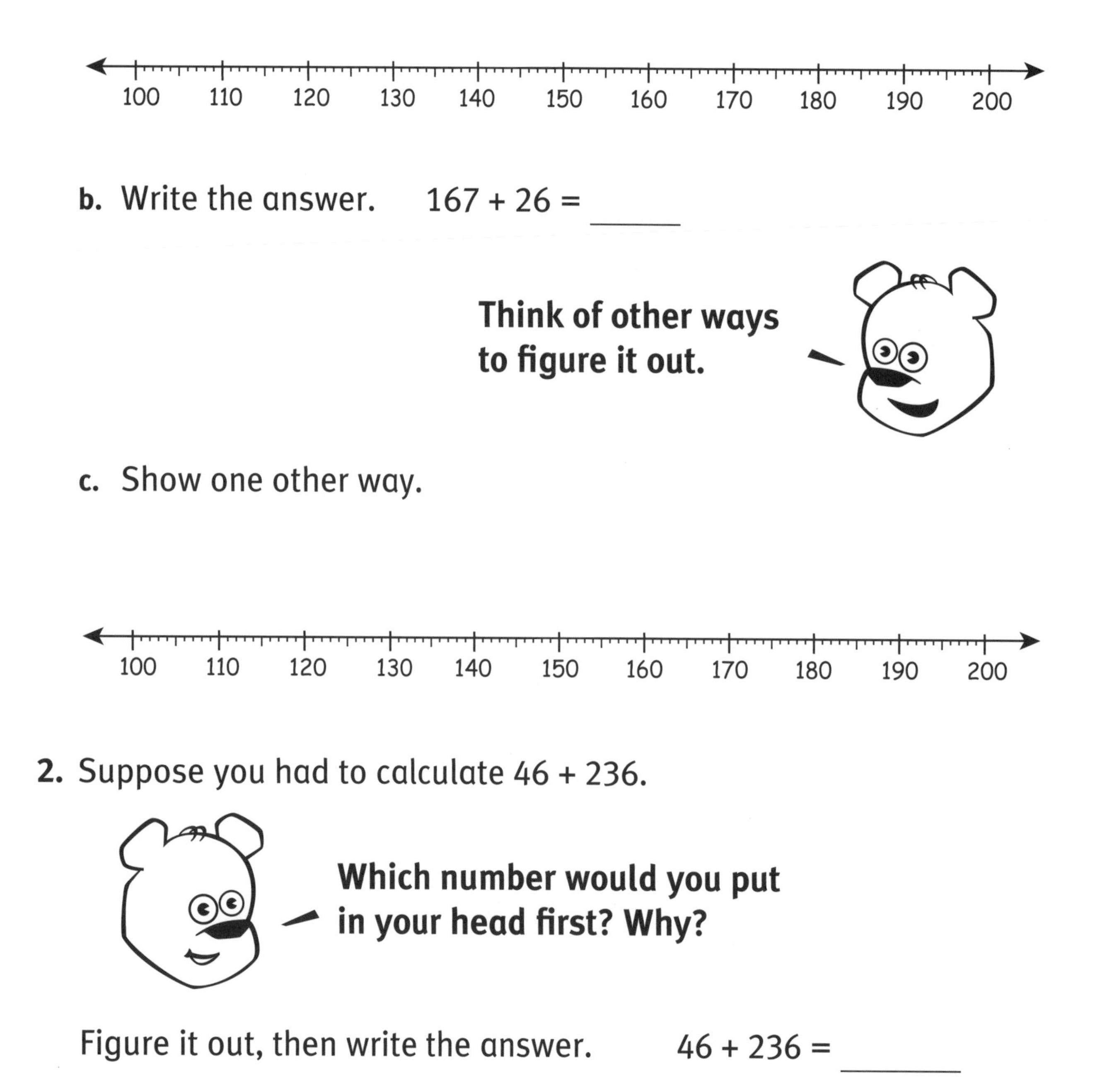

b. Write the answer. 167 + 26 = ______

Think of other ways to figure it out.

c. Show one other way.

100 110 120 130 140 150 160 170 180 190 200

2. Suppose you had to calculate 46 + 236.

Which number would you put in your head first? Why?

Figure it out, then write the answer. 46 + 236 = ________

Name: ______________________

1. Figure out the total of each of these in your head. Write the answer. Draw jumps to show how you did it.

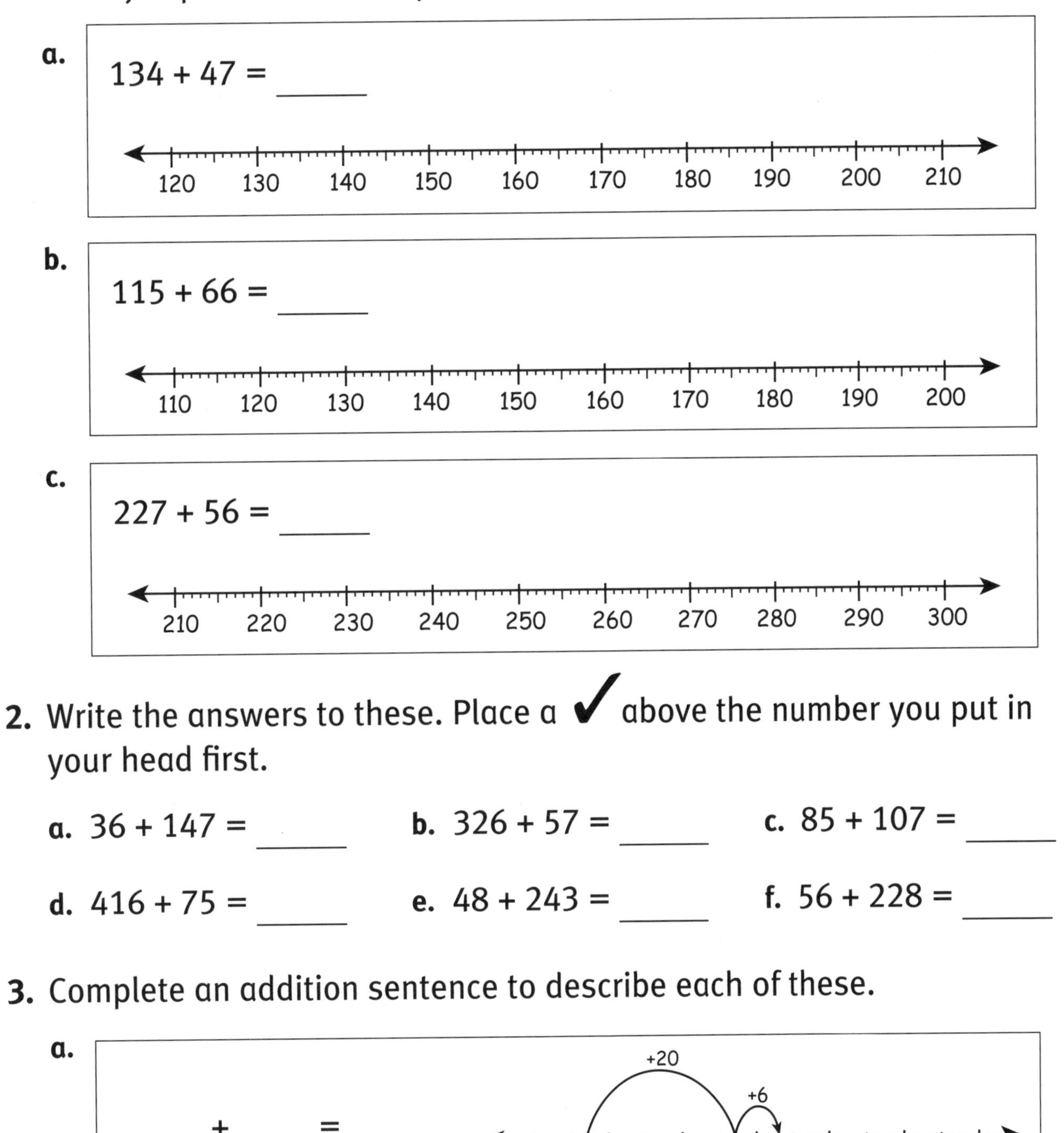

a. 134 + 47 = ______

b. 115 + 66 = ______

c. 227 + 56 = ______

2. Write the answers to these. Place a ✓ above the number you put in your head first.

a. 36 + 147 = ______ **b.** 326 + 57 = ______ **c.** 85 + 107 = ______

d. 416 + 75 = ______ **e.** 48 + 243 = ______ **f.** 56 + 228 = ______

3. Complete an addition sentence to describe each of these.

a. ______ + ______ = ______

+20
+6
247

b. ______ + ______ = ______

+10
+6
378

Name: ______________________

Jenny needs to ride to the store and back. It is 6.4 miles each way. What is the total distance?

1. a. Figure out the answer in your head. Write the number sentence.

_____ + _____ = _____

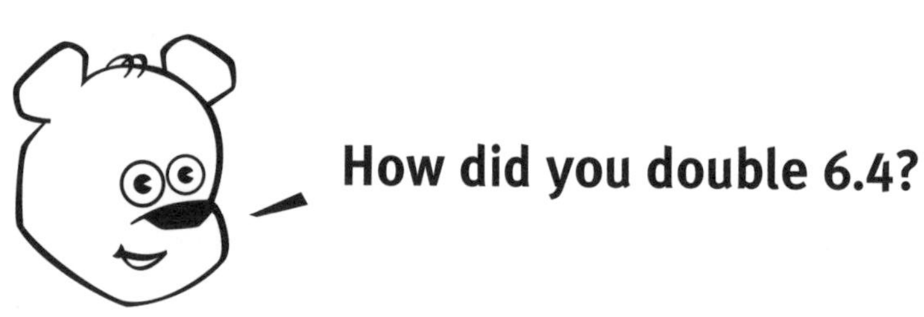

b. Complete this sentence.

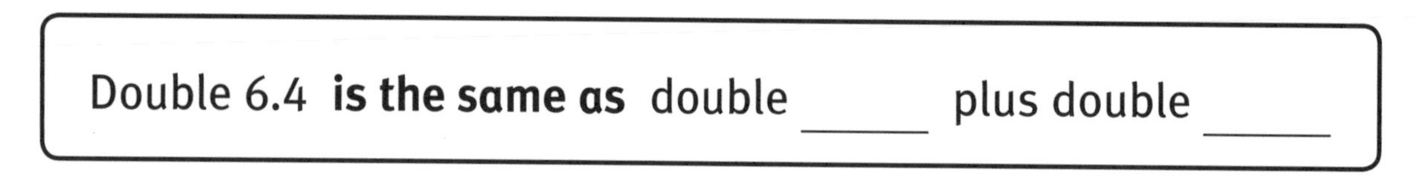

Double 6.4 **is the same as** double _____ plus double _____

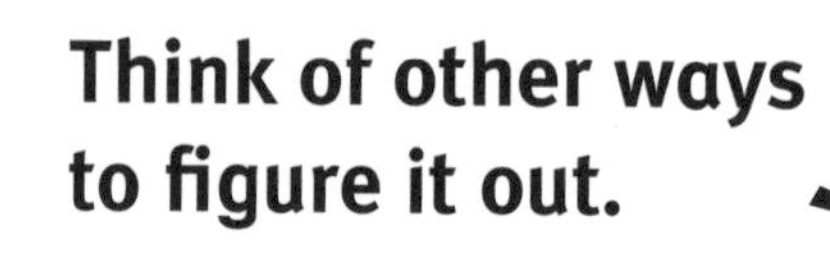

c. Complete this sentence to show another method.

Double 6.4 **is the same as** ______________________

2. Suppose it was 4.3 miles each way.

Complete this sentence to show how you could figure out the total distance to the store and back.

Double 4.3 **is the same as** ______________________

Name: ______________________

1. Suppose you had to figure out 21.4 + 21.4 in your head.
Write how you would do it.

____

____

2. For each of these, complete the sentence to show one way you could figure it out. Write the answer.

a. Double 31.2 **is the same as**

____

____

31.2 + 31.2 = ______

b. Double 25.4 **is the same as**

____

____

25.4 + 25.4 = ______

c. Double 12.3 **is the same as**

____

____

12.3 + 12.3 = ______

d. Double 45.1 **is the same as**

____

____

45.1 + 45.1 = ______

3. Double each of these. Write the number sentence.

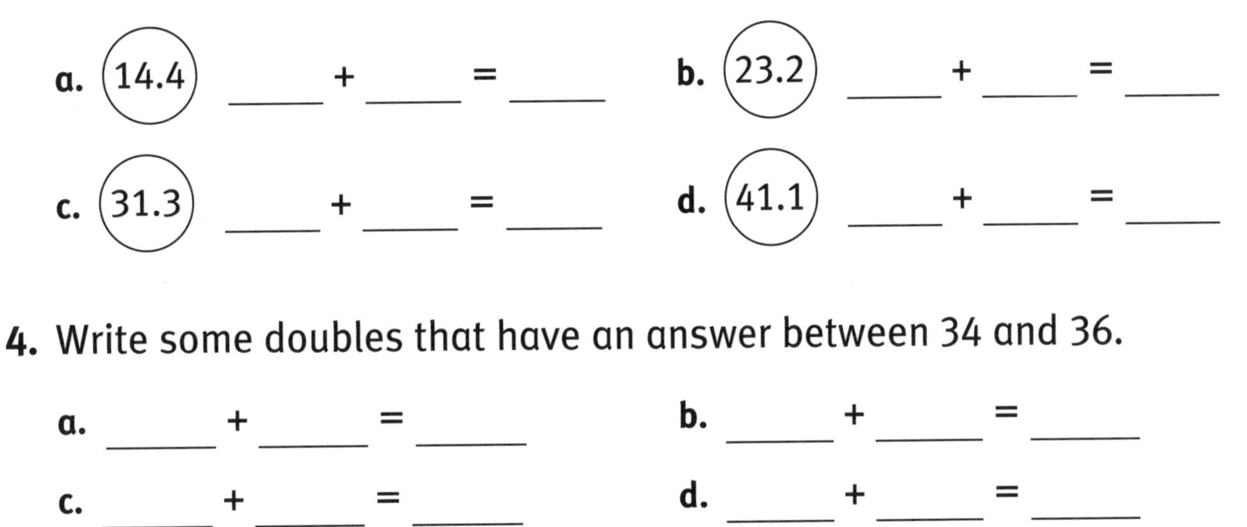

a. 14.4 ______ + ______ = ______

b. 23.2 ______ + ______ = ______

c. 31.3 ______ + ______ = ______

d. 41.1 ______ + ______ = ______

4. Write some doubles that have an answer between 34 and 36.

a. ______ + ______ = ______

b. ______ + ______ = ______

c. ______ + ______ = ______

d. ______ + ______ = ______

WARM UP 3

Name: ____________________

Tyson bought a pair of soccer shoes for $36 and socks for $6.23. What was the total cost?

The Soccer Store

1. Tyson put one number in his head then added the parts of the other number like this.

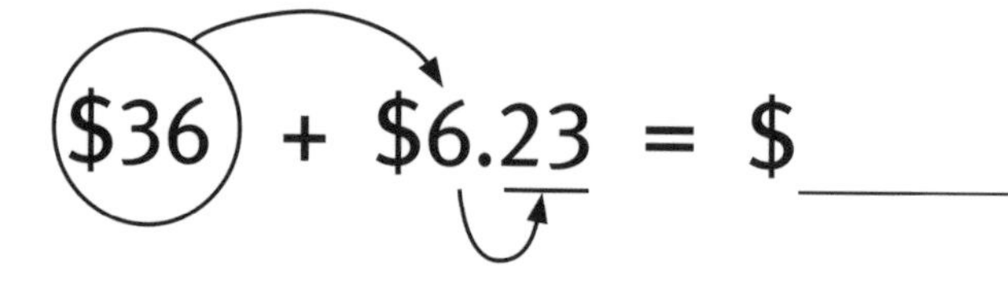

Write the answer above.

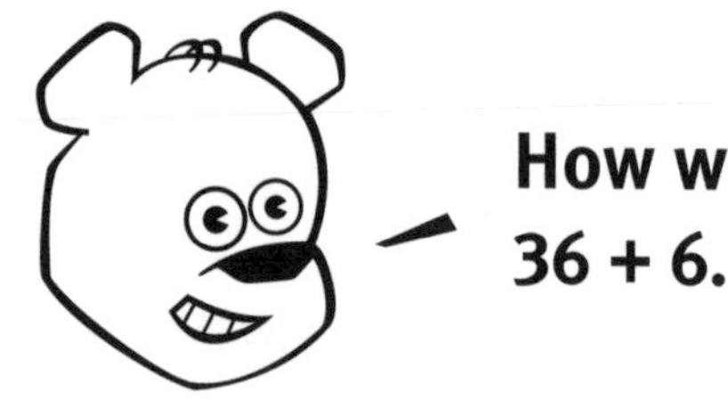

How would you figure out 36 + 6.23 in your head?

2. Suppose the shoes were $48 and the socks were $5.82. What would be the total cost?

Try putting one amount in your head then adding the parts of the other number.

Write the number sentence. Draw arrows to show how you added.

______ + ______ = ______

Name: ____________________

1. Look at this number sentence. Write how you know it is correct.

23 + 4.72 = 27.72

____

____

2. For each of these, follow the arrows to add the parts of the second number in your head. Write the answer.

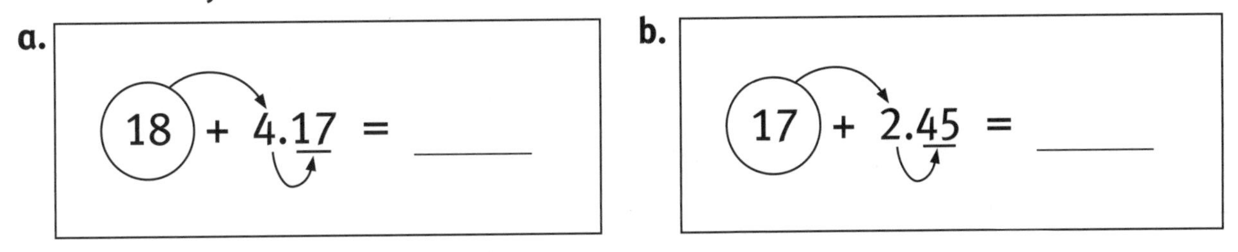

3. Ring the number you would put in your head first. Add the parts of the other number. Write the answer.

a. 36 + 5.32 = ______

b. 6.47 + 22 = ______

c. 3.28 + 26 = ______

d. 49 + 3.33 = ______

4. For each of these, roll a number cube and write the number in the box. Add the numbers and write the answer.

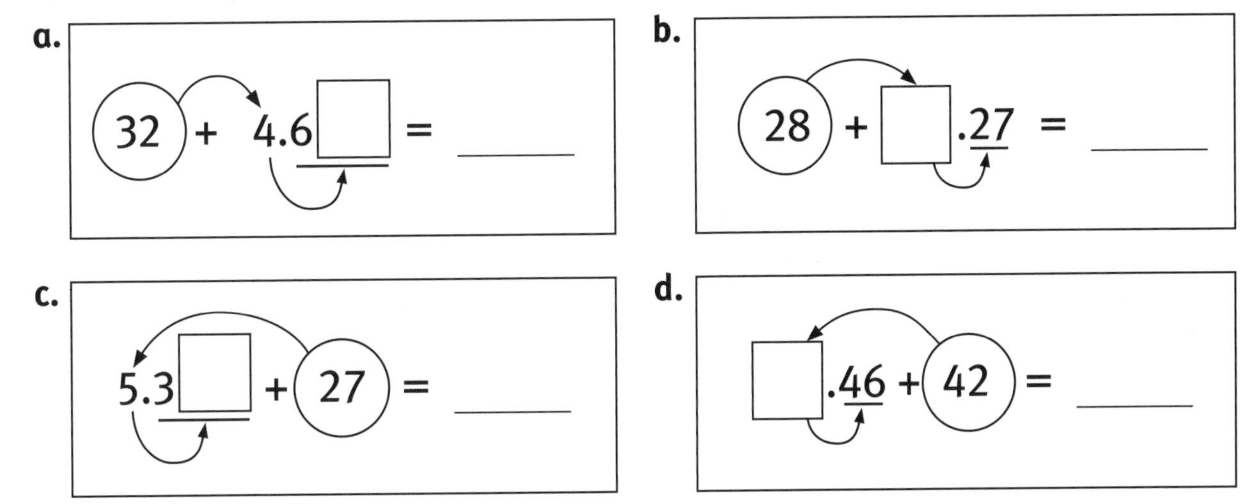

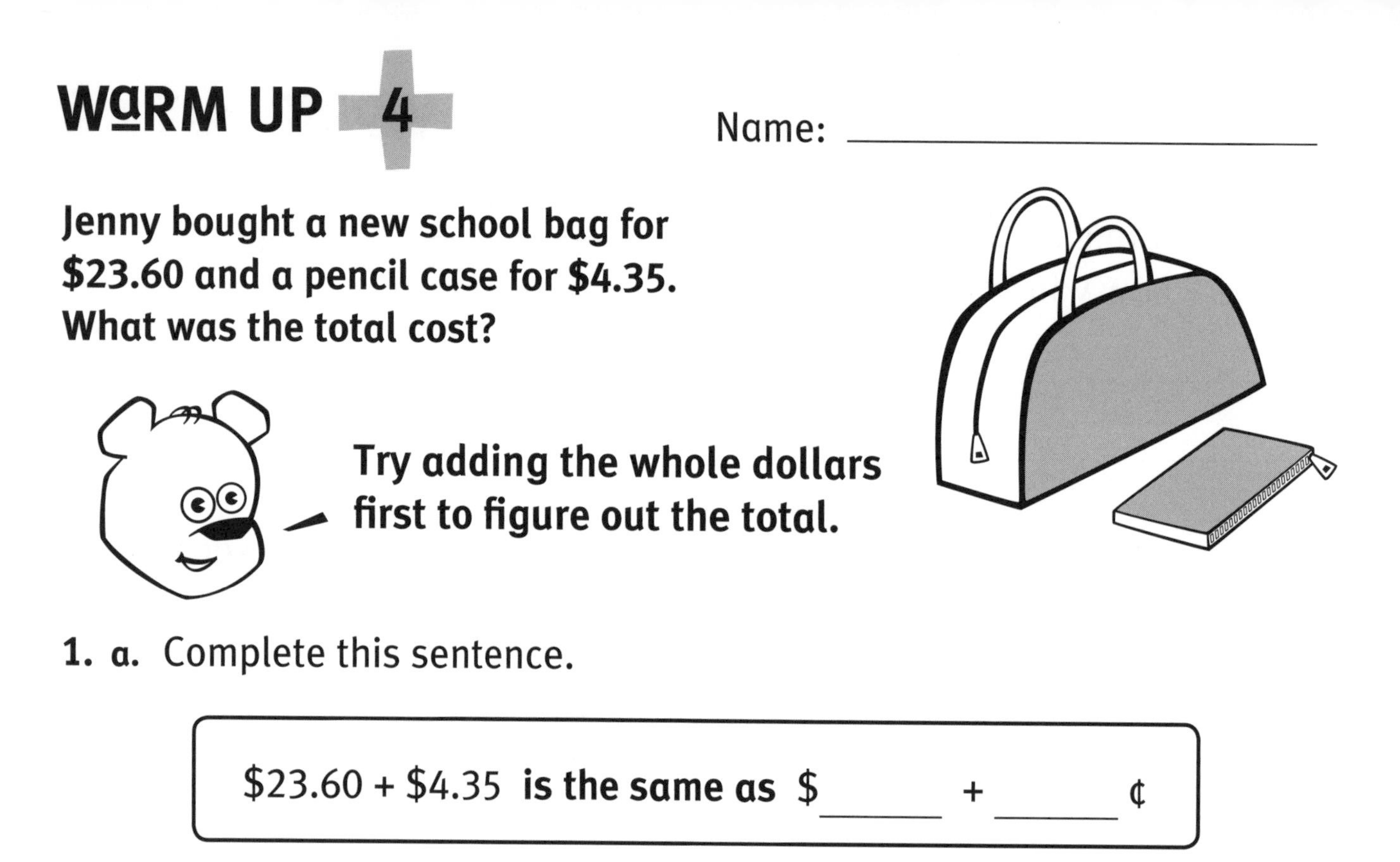

WARM UP 4

Name: ______________________

Jenny bought a new school bag for $23.60 and a pencil case for $4.35. What was the total cost?

Try adding the whole dollars first to figure out the total.

1. a. Complete this sentence.

   $23.60 + $4.35 **is the same as** $_____ + _____¢

   b. Write the total. ________

Describe another way you could figure out the total.

2. Suppose the bag was $32.25 and the pencil case was $4.55.

Figure out the total cost by adding the dollars then the cents.

   a. Complete this sentence.

   $32.25 + $4.55 **is the same as** $_____ + _____¢

   b. Write the total. ________

Name: ______________________

1. Jenny wanted to figure out $34.20 + $3.05. She thought the total was $37.70. How can you tell she made a mistake?

____

____

2. For each of these, complete the sentence then write the answer.

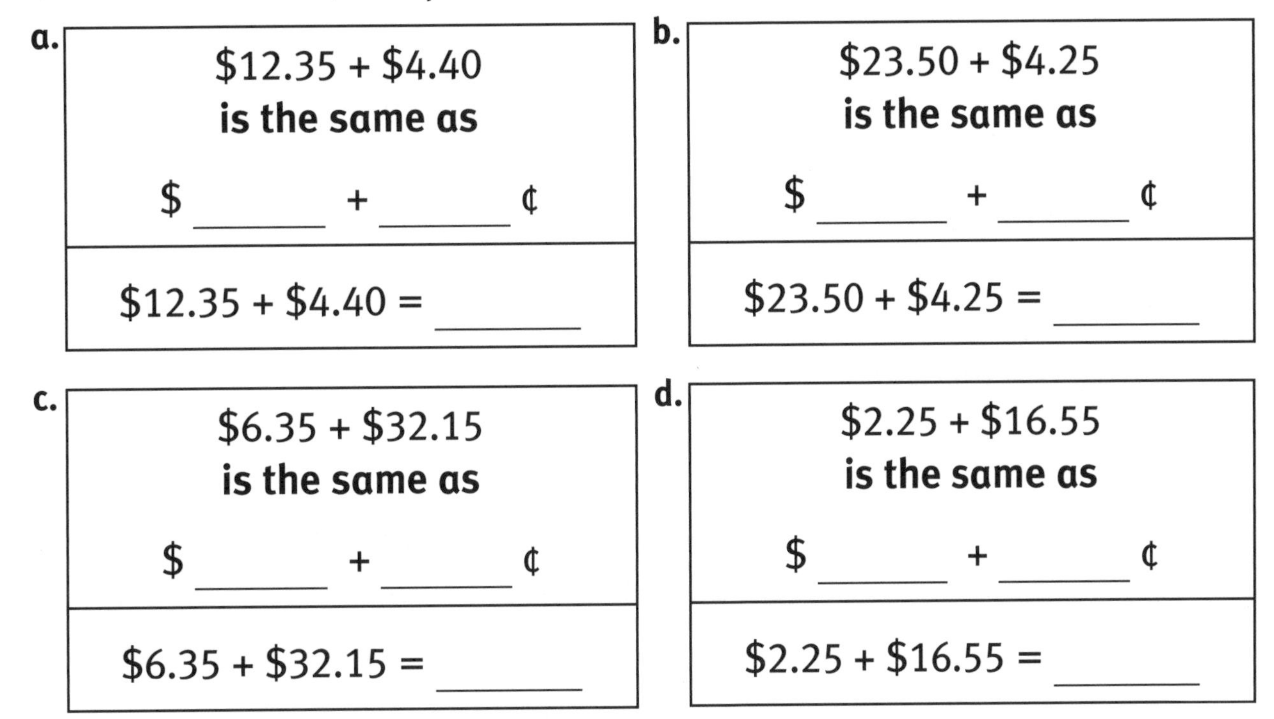

a.
$12.35 + $4.40
is the same as
$ ______ + ______ ¢

$12.35 + $4.40 = ______

b.
$23.50 + $4.25
is the same as
$ ______ + ______ ¢

$23.50 + $4.25 = ______

c.
$6.35 + $32.15
is the same as
$ ______ + ______ ¢

$6.35 + $32.15 = ______

d.
$2.25 + $16.55
is the same as
$ ______ + ______ ¢

$2.25 + $16.55 = ______

3. Write the answers. Try adding the dollars then the cents.

a. $42.15 + $3.50 = ______ **b.** $4.25 + $12.60 = ______

c. $23.40 + $3.35 = ______ **d.** $12.20 + $12.35 = ______

4. Write some number sentences involving dollars and cents that you could solve the same way.

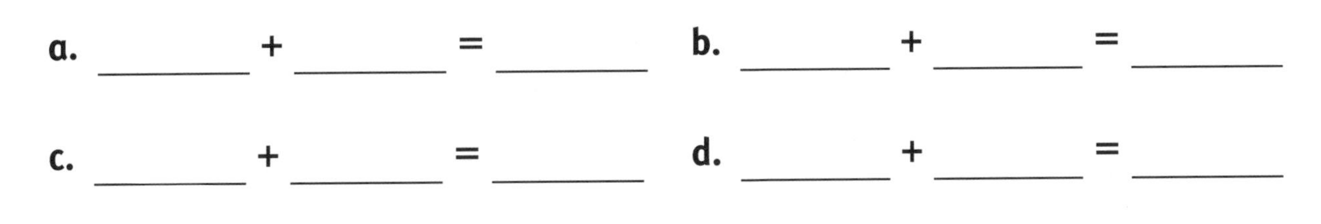

a. ______ + ______ = ______ **b.** ______ + ______ = ______

c. ______ + ______ = ______ **d.** ______ + ______ = ______

Name: ______________________

A car rally was 7.8 miles to the river then 3.6 miles to the finish. How far was the race?

Try rounding one or both numbers to help you figure out the answer.

1. a. Complete this sentence to describe what you did.

7.8 + 3.6 **is the same as** ______________________

b. Write the answer. ______

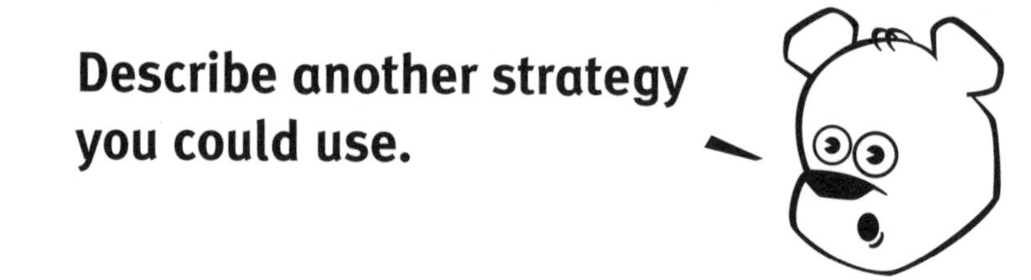

c. Complete this sentence to show another method.

7.8 + 3.6 **is the same as** ______________________

2. Suppose you had to figure out 6.9 + 2.4.
Write an easier number sentence that would help.
Write the answer.

______ + ______ = ______ **so** 6.9 + 2.4 = ______

Name: ______________________

WORK OUT 5

1. For each of these, write an easier number sentence that will help you figure out the problem below. Write the answer.

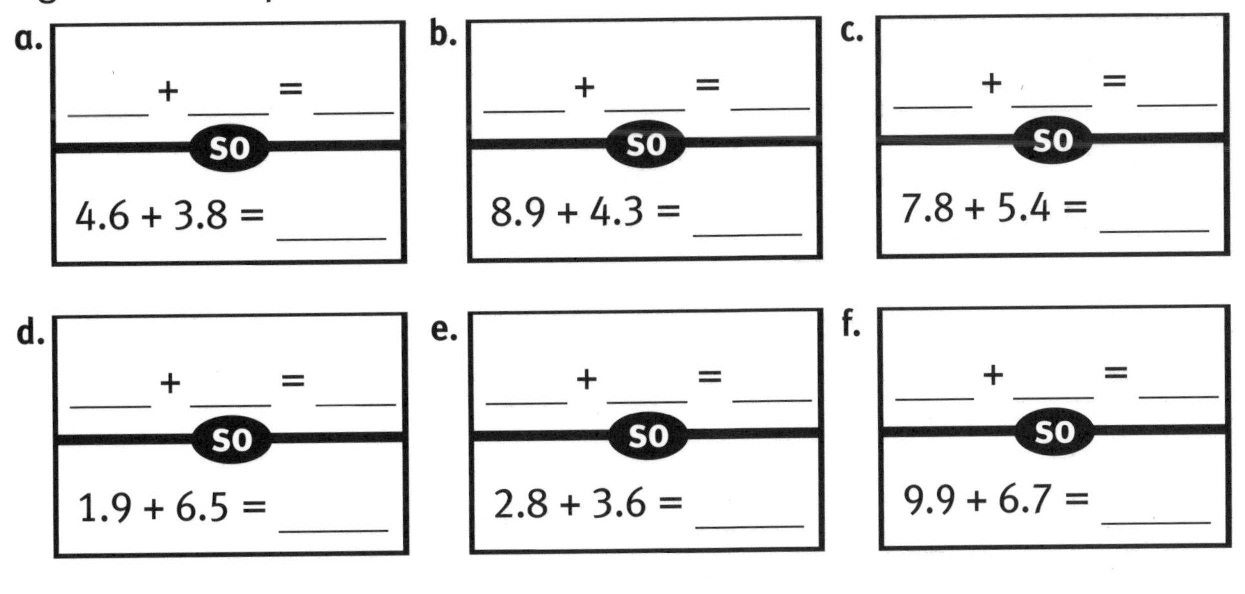

2. Write the answers. Place a ✔ above the numbers you adjusted.

a. 6.9 + 8.4 = ______ **b.** 5.3 + 7.8 = ______ **c.** 8.9 + 4.4 = ______

d. 7.7 + 8.8 = ______ **e.** 6.5 + 3.9 = ______ **f.** 4.8 + 4.9 = ______

3. Add the numbers on the spokes to the number in the center. Write the answers around the outside.

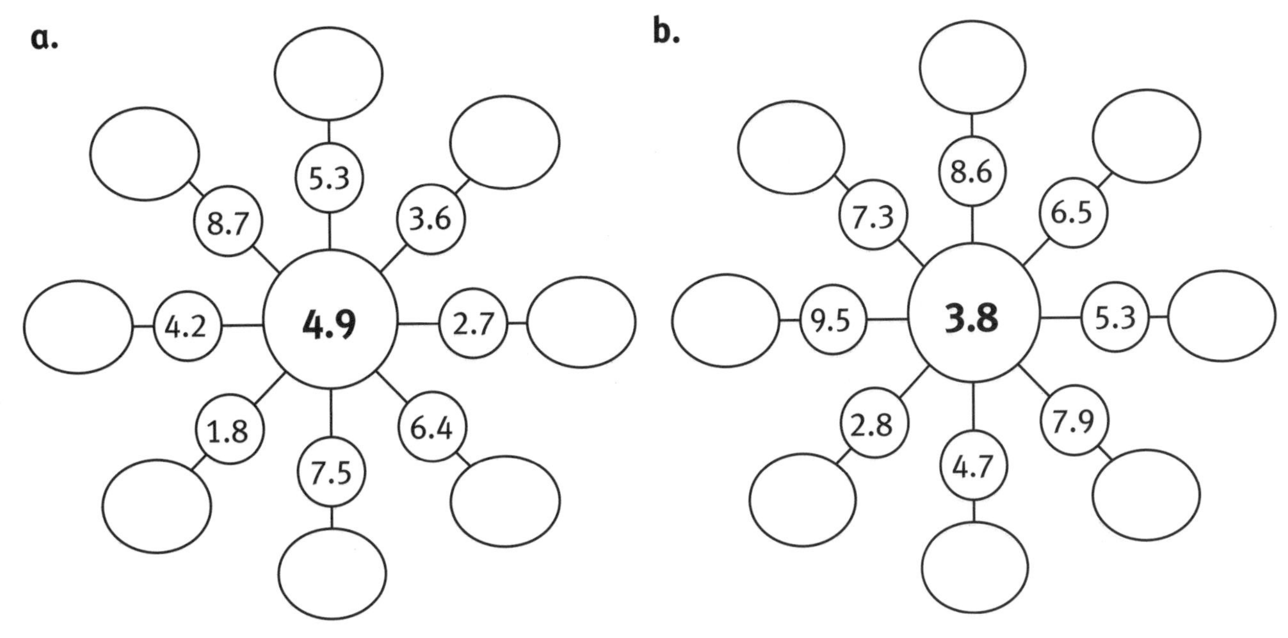

WaRM UP 6

Name: ______________________

Kelly and Anita bought two gifts that cost $11.95 and $13.95. What was the total cost?

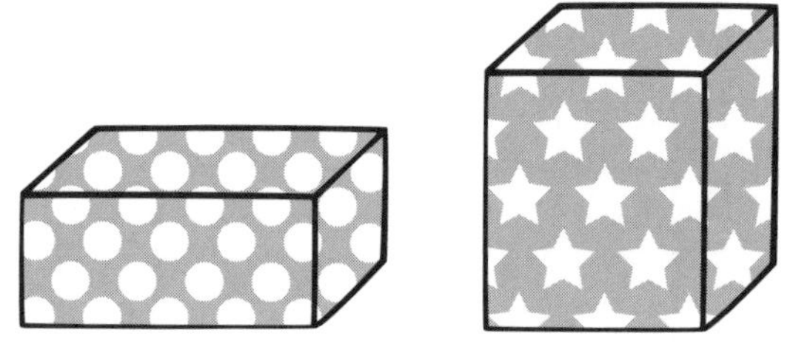

1. Kelly rounded both numbers to help her.
   Complete this sentence to show how she may have figured it out.

$11.95 + $13.95 **is the same as** ______________________

2. Anita took 5 cents from one price and added it to the other amount.
   Use Anita's method to write an easier number sentence.

$11.95 + $13.95 **is the same as** ______________________

Think of another way to calculate the total cost.

3. Which method do you like best? Why?

____

____

4. Write a number sentence you could solve using the method you prefer.

_______ + _______ = _______

Name: ____________________

1. a. Suppose you had to figure out $19.95 + $4.95 in your head. Write about the strategy you would use.

____

____

b. Write two other problems with dollars and cents that you could solve using this method.

_______ + _______ = _______   _______ + _______ = _______

2. Use the method you prefer to calculate each of these.

a. $9.95 + $5.95 = _______

b. $8.98 + $7.98 = _______

c. $16.99 + $2.99 = _______

d. $12.95 + $4.99 = _______

e. $23.98 + $5.95 = _______

f. $34.99 + $4.98 = _______

g. $33.99 + $4.95 = _______

h. $42.98 + $6.98 = _______

3. For each of these, draw an arrow to a number sentence you could use to help figure out the total. Write the answers.

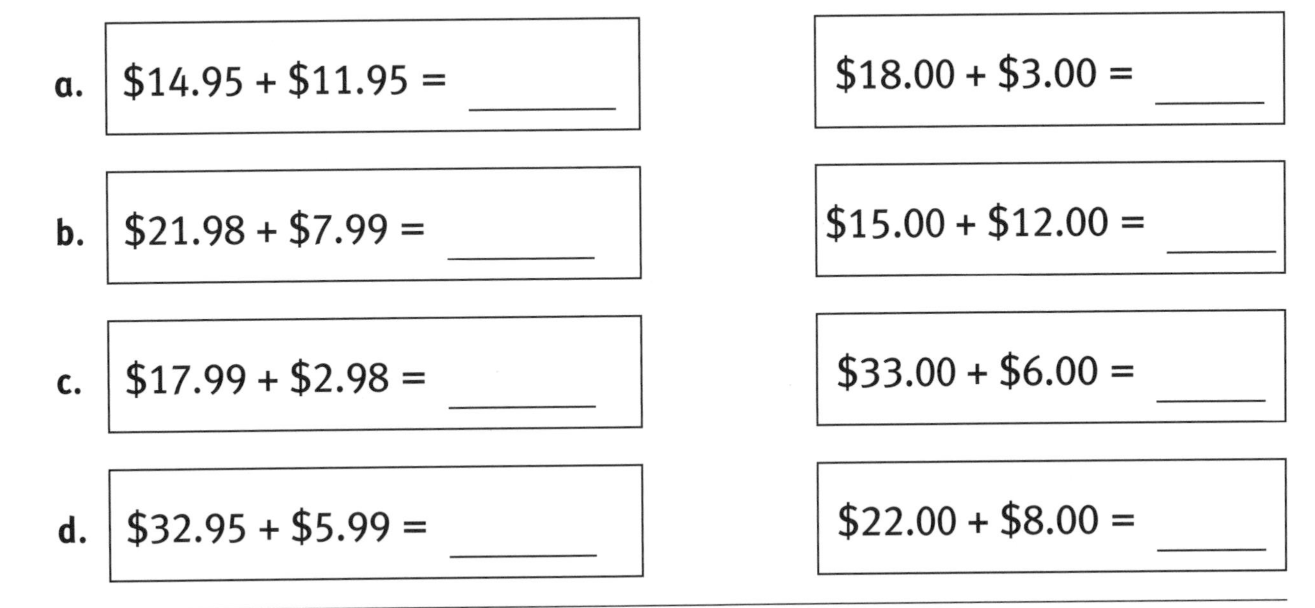

a. $14.95 + $11.95 = _______

b. $21.98 + $7.99 = _______

c. $17.99 + $2.98 = _______

d. $32.95 + $5.99 = _______

$18.00 + $3.00 = _______

$15.00 + $12.00 = _______

$33.00 + $6.00 = _______

$22.00 + $8.00 = _______

WarM UP 7

Name: ______________________

Friends had a pizza party. There was $\frac{1}{2}$ of one pizza left over and $\frac{3}{4}$ of another. How much pizza was left over?

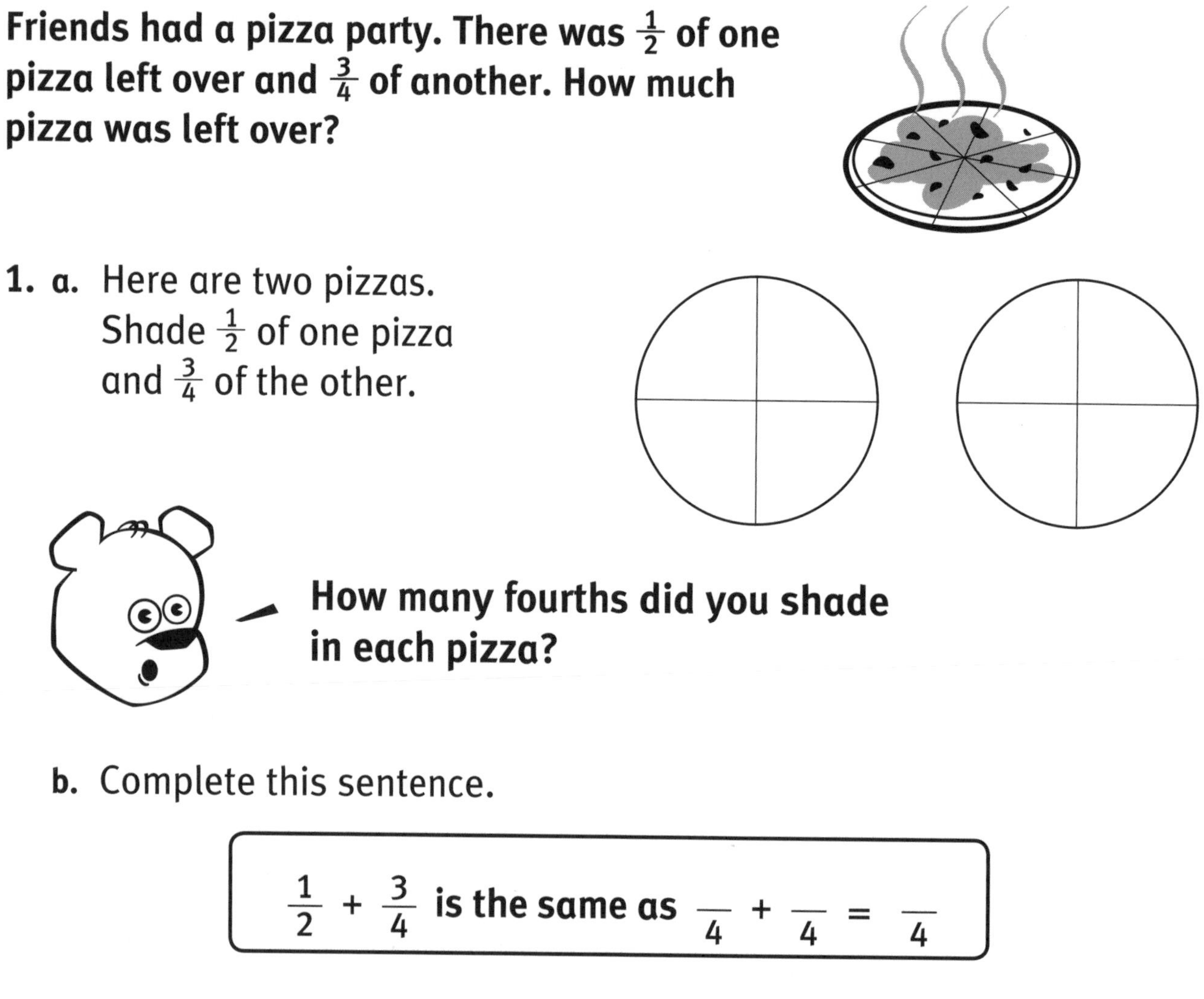

1. a. Here are two pizzas. Shade $\frac{1}{2}$ of one pizza and $\frac{3}{4}$ of the other.

How many fourths did you shade in each pizza?

b. Complete this sentence.

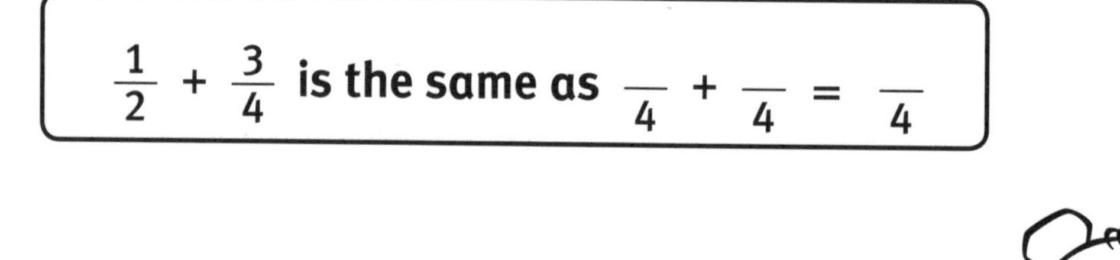

$\frac{1}{2} + \frac{3}{4}$ **is the same as** $\frac{\_}{4} + \frac{\_}{4} = \frac{\_}{4}$

How much pizza was left over?

2. Suppose $\frac{1}{2}$ of one pizza and $\frac{5}{8}$ of the other were left over. Shade these pizzas to show each fraction. Complete the sentence.

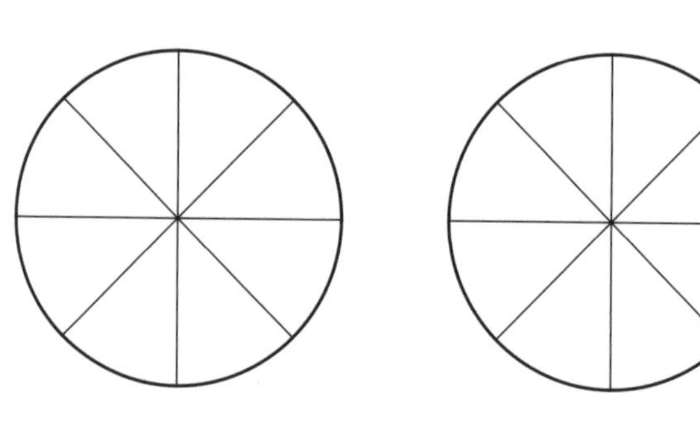

$\frac{1}{2} + \frac{5}{8}$ **is the same as** $\frac{\_}{8} + \frac{\_}{8} = \frac{\_}{8}$

Name: ______________________

1. For each of these, shade each fraction then write an equivalent number sentence. Write the answer.

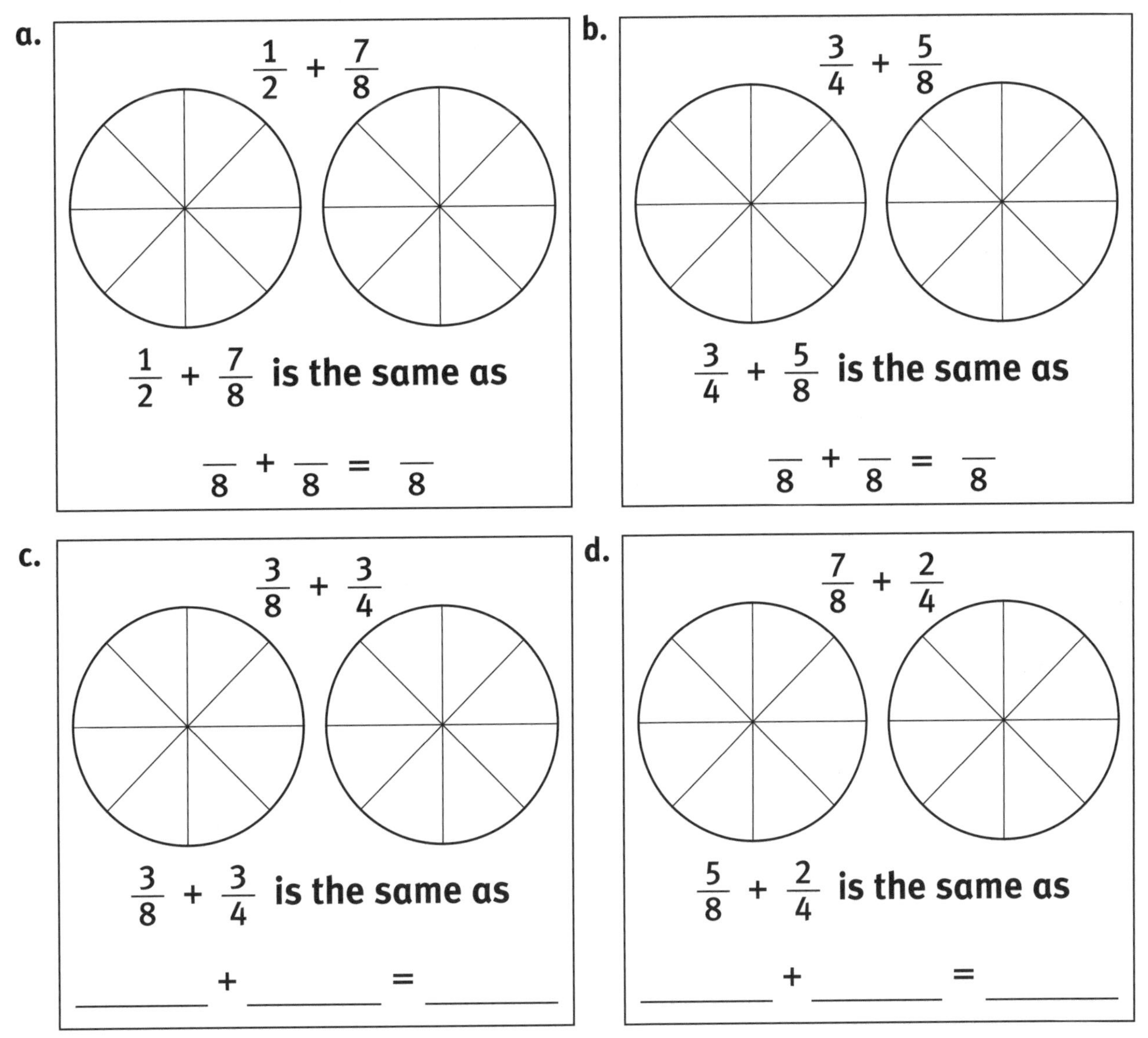

2. Draw arrows from each of these to the correct answer below. Use equivalent fractions to help you.

a. $\frac{3}{4} + \frac{7}{8}$ **b.** $\frac{1}{2} + \frac{3}{8}$ **c.** $\frac{1}{4} + \frac{7}{8}$ **d.** $\frac{5}{8} + \frac{3}{4}$

$\frac{9}{8}$ or $1\frac{1}{8}$ $\frac{11}{8}$ or $1\frac{3}{8}$ $\frac{13}{8}$ or $1\frac{5}{8}$ $\frac{7}{8}$

Name: ______________________

A boat sailed 1.9, 2.7, 3.3, and 4.1 nautical miles between several islands. How far did it sail in all?

Look for an easy way to figure out the answer.

1. Write a number sentence to show the order in which you would add the numbers. Write the answer.

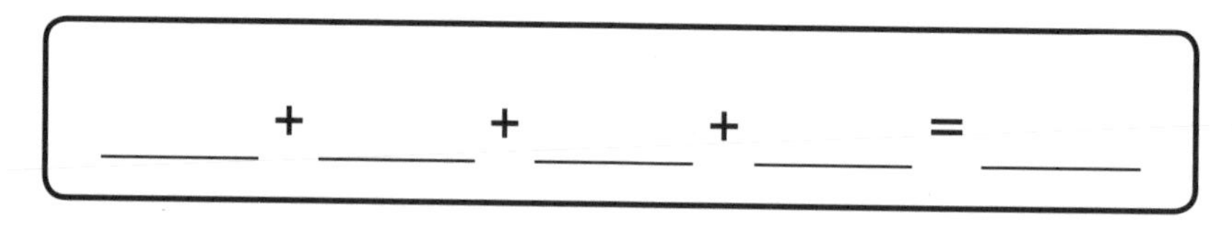

2. Suppose the boat sailed 6.4, 3.8, 1.6, and 4.2 nautical miles.

Look for pairs that add nicely together.

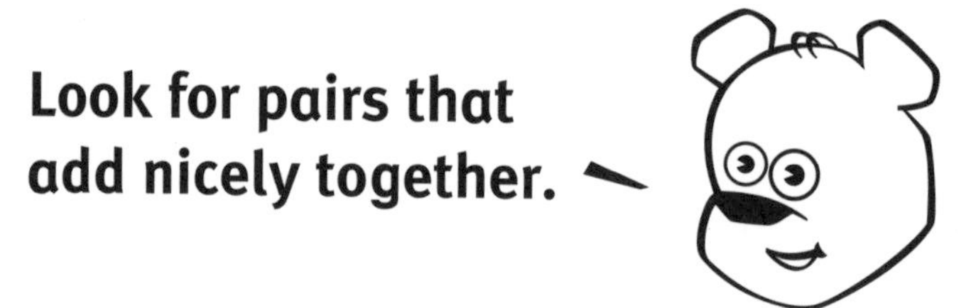

a. Write the numbers in the order you would add them to figure out the total distance in your head.

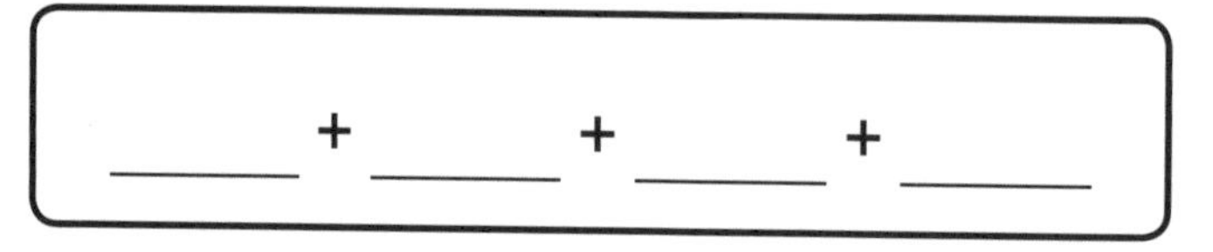

b. Write the total distance. ______ nautical miles

Name: ______________________

WORK OUT 8

1. For each of these, ring pairs of numbers that add nicely together. Figure out and write the total.

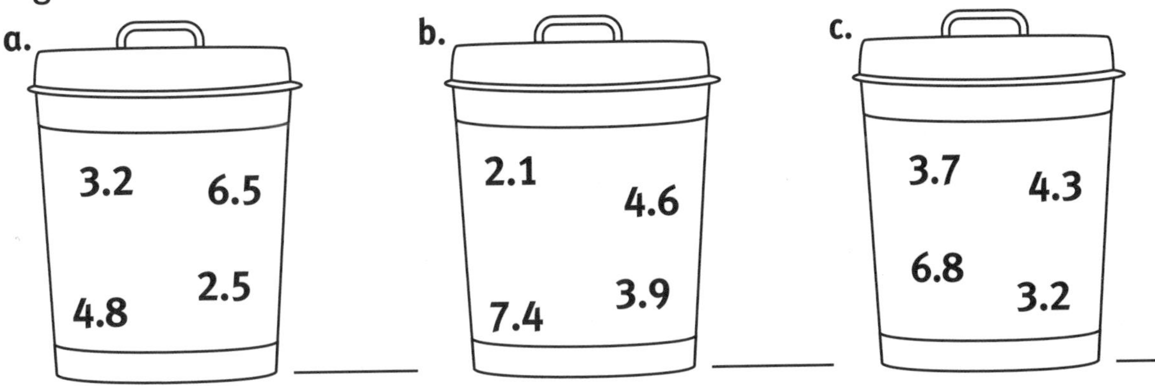

2. Figure out the distance the boat sailed on each day. Write the total.

					Total
Monday	4.3	6.2	7.7	1.8	
Tuesday	8.5	3.6	3.4	7.5	
Wednesday	5.9	7.1	0.6	5.4	
Thursday	11.2	4.3	7.8	0	
Friday	2.7	0	4.6	0.3	

3. Draw ⌒ to connect numbers you would add together first. Write the answer.

a. 12.6 + 3.5 + 12.4 + 7.5 = ________

b. 8.7 + 9.1 + 12.3 + 6.9 = ________

c. 24.2 + 12.1 + 7.9 + 13.8 = ________

d. 20.3 + 4.7 + 8.6 + 11.4 = ________

CHECK UP 1

Name: ____________________

1. Figure out these in your head. Write the answers.

a. 168 + 37 = ______ **b.** 21.2 + 21.2 = ______ **c.** 4.9 + 3.5 = ______

d. $3.25 + $12.55 = ________ **e.** $13.95 + $3.99 = ________

2. a. Write the answer.

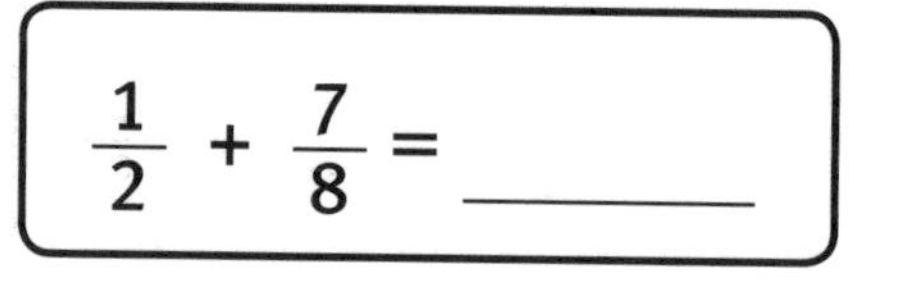

$\frac{1}{2} + \frac{7}{8} =$ ________

b. Write how you figured it out in your head.

____

____

c. Write two other problems with fractions that you could solve the same way.

_____ + _____ = _____ _____ + _____ = _____

3. For each of these, write the answer then write a similar number sentence you could solve the same way.

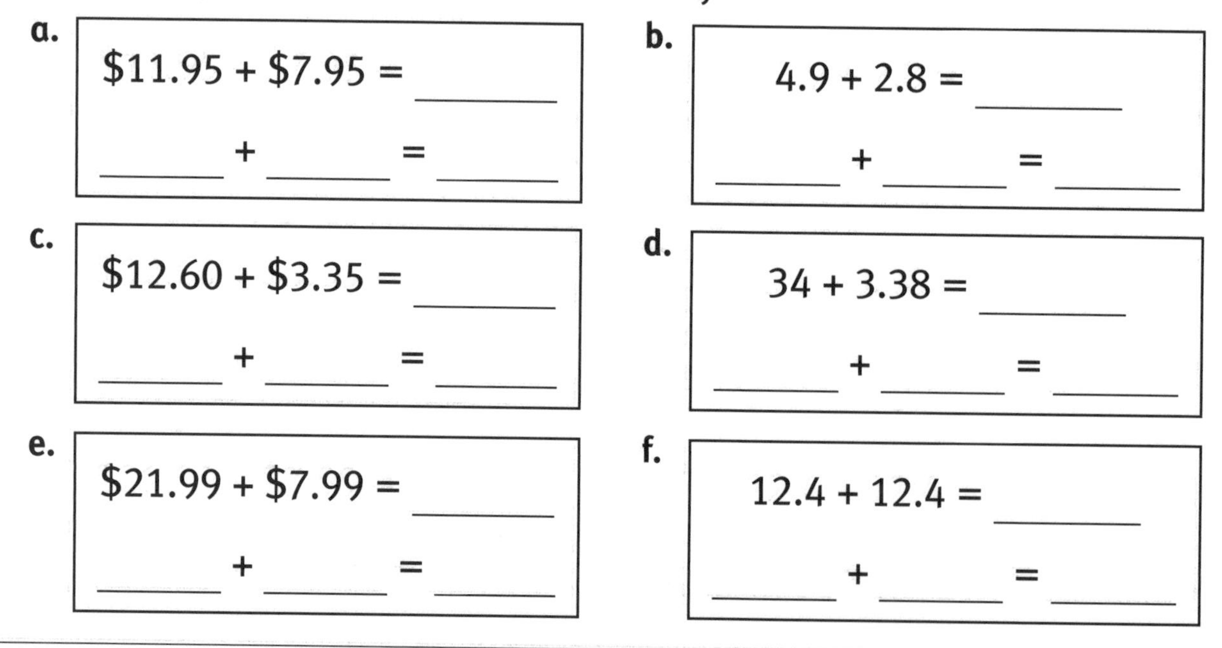

a. $11.95 + $7.95 = ______
_____ + _____ = _____

b. 4.9 + 2.8 = ______
_____ + _____ = _____

c. $12.60 + $3.35 = ______
_____ + _____ = _____

d. 34 + 3.38 = ______
_____ + _____ = _____

e. $21.99 + $7.99 = ______
_____ + _____ = _____

f. 12.4 + 12.4 = ______
_____ + _____ = _____

Name: ____________________

Suppose you were in a race and you passed the person in second place. What place would you be in?

Solve this riddle by figuring out the answers below. Write the letters above their matching answers at the bottom of the page. Some letters appear more than once. The first one is done for you.

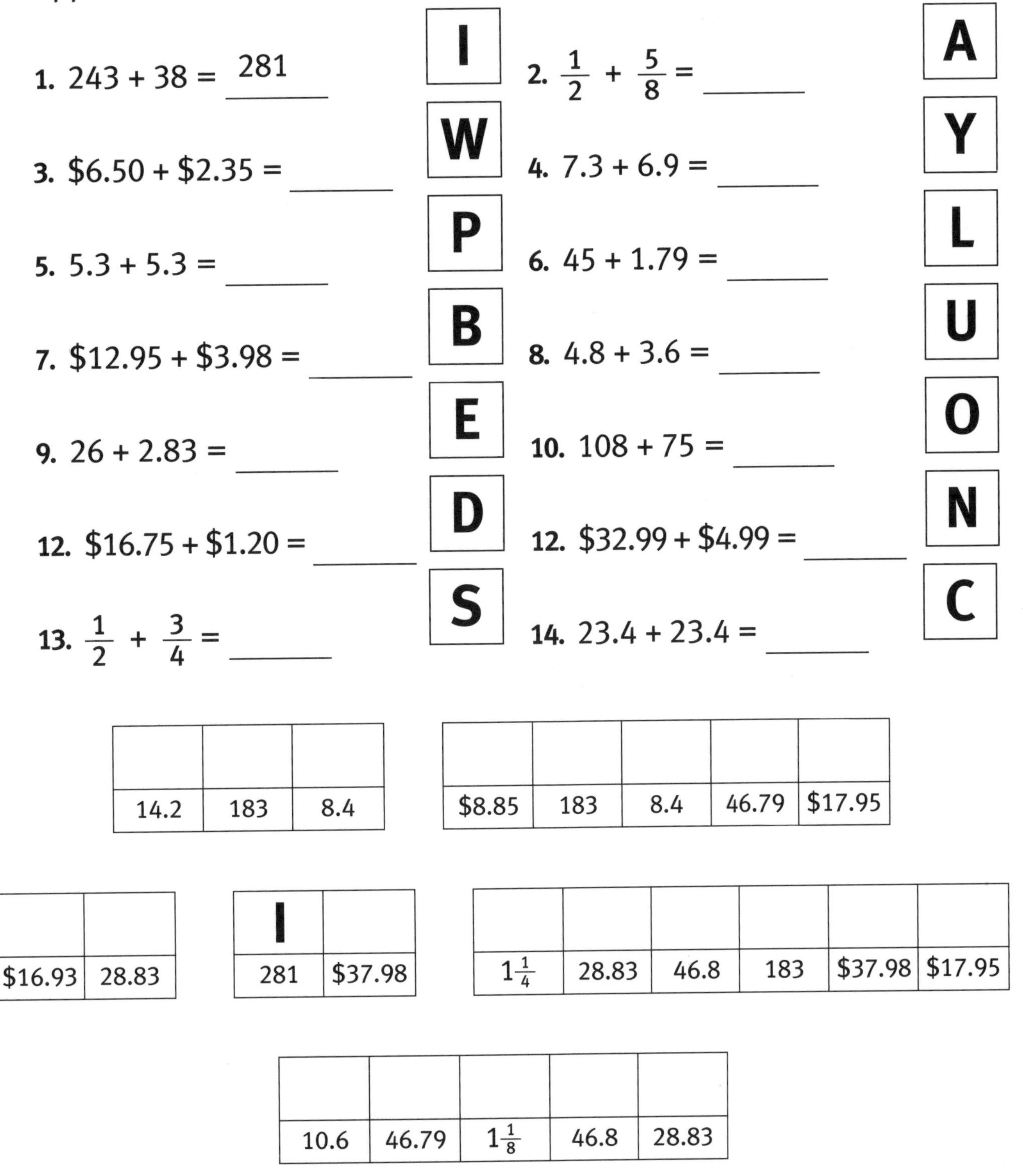

1. $243 + 38 =$ 281 **I**
2. $\frac{1}{2} + \frac{5}{8} =$ ______ **A**
3. $6.50 + $2.35 = ______ **W**
4. $7.3 + 6.9 =$ ______ **Y**
5. $5.3 + 5.3 =$ ______ **P**
6. $45 + 1.79 =$ ______ **L**
7. $12.95 + $3.98 = ______ **B**
8. $4.8 + 3.6 =$ ______ **U**
9. $26 + 2.83 =$ ______ **E**
10. $108 + 75 =$ ______ **O**
12. $16.75 + $1.20 = ______ **D**
12. $32.99 + $4.99 = ______ **N**
13. $\frac{1}{2} + \frac{3}{4} =$ ______ **S**
14. $23.4 + 23.4 =$ ______ **C**

14.2	183	8.4

$8.85	183	8.4	46.79	$17.95

$16.93	28.83

I	
281	$37.98

$1\frac{1}{4}$	28.83	46.8	183	$37.98	$17.95

10.6	46.79	$1\frac{1}{8}$	46.8	28.83

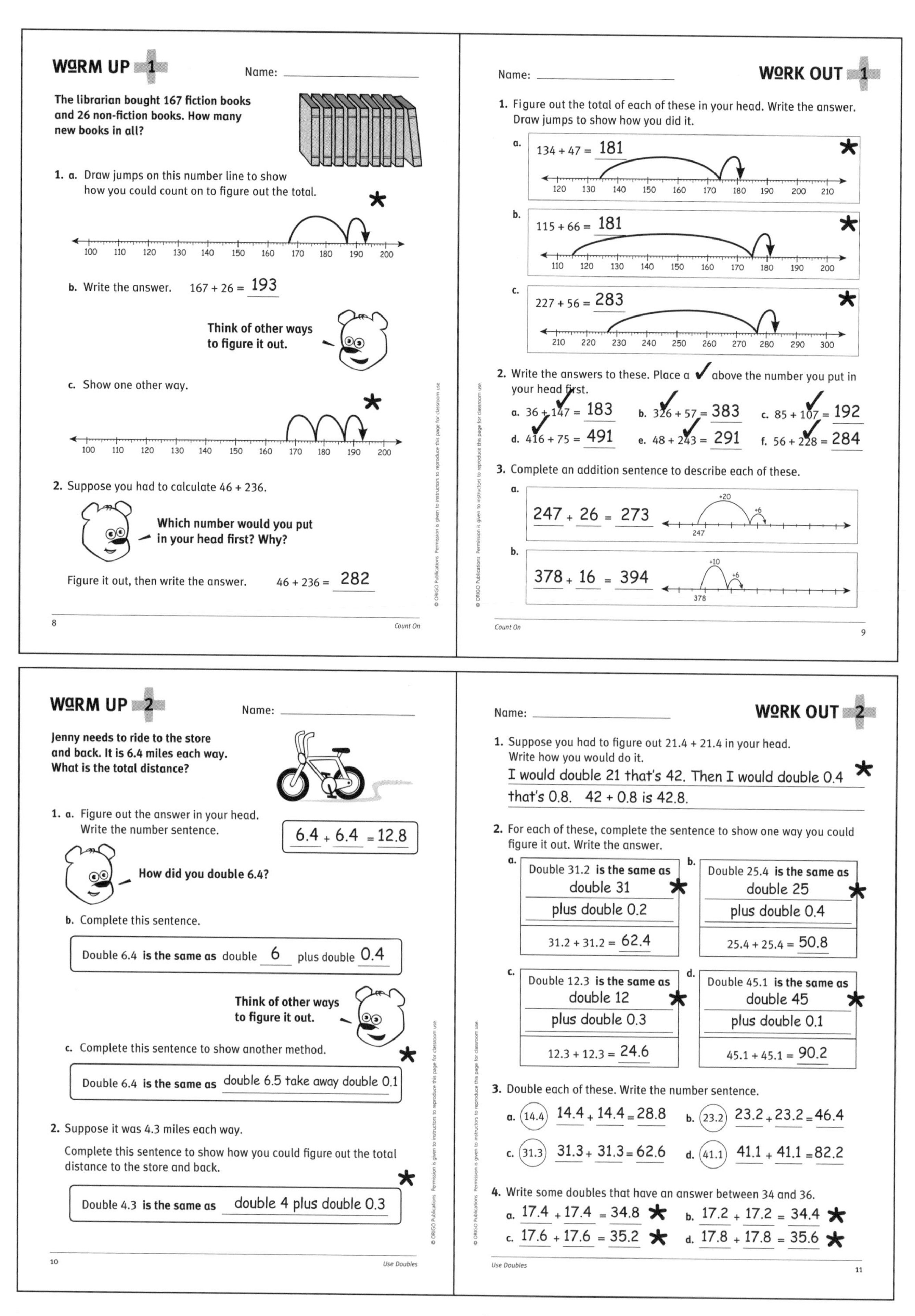

WORM UP 1

Name: ____________

The librarian bought 167 fiction books and 26 non-fiction books. How many new books in all?

1. a. Draw jumps on this number line to show how you could count on to figure out the total. *

100 110 120 130 140 150 160 170 180 190 200

b. Write the answer. 167 + 26 = 193

Think of other ways to figure it out.

c. Show one other way. *

100 110 120 130 140 150 160 170 180 190 200

2. Suppose you had to calculate 46 + 236.

Which number would you put in your head first? Why?

Figure it out, then write the answer. 46 + 236 = 282

8 Count On

WORK OUT 1

Name: ____________

1. Figure out the total of each of these in your head. Write the answer. Draw jumps to show how you did it.

a. 134 + 47 = 181 *

120 130 140 150 160 170 180 190 200 210

b. 115 + 66 = 181 *

110 120 130 140 150 160 170 180 190 200

c. 227 + 56 = 283 *

210 220 230 240 250 260 270 280 290 300

2. Write the answers to these. Place a ✓ above the number you put in your head first.

a. 36 + 147 = 183 b. 326 + 57 = 383 c. 85 + 107 = 192

d. 416 + 75 = 491 e. 48 + 243 = 291 f. 56 + 228 = 284

3. Complete an addition sentence to describe each of these.

a. 247 + 26 = 273 (+20, +6, 247)

b. 378 + 16 = 394 (+10, +6, 378)

Count On 9

WORM UP 2

Name: ____________

Jenny needs to ride to the store and back. It is 6.4 miles each way. What is the total distance?

1. a. Figure out the answer in your head. Write the number sentence. 6.4 + 6.4 = 12.8

How did you double 6.4?

b. Complete this sentence.

Double 6.4 is the same as double 6 plus double 0.4

Think of other ways to figure it out.

c. Complete this sentence to show another method. *

Double 6.4 is the same as double 6.5 take away double 0.1

2. Suppose it was 4.3 miles each way.

Complete this sentence to show how you could figure out the total distance to the store and back. *

Double 4.3 is the same as double 4 plus double 0.3

10 Use Doubles

WORK OUT 2

Name: ____________

1. Suppose you had to figure out 21.4 + 21.4 in your head. Write how you would do it.

I would double 21 that's 42. Then I would double 0.4 that's 0.8. 42 + 0.8 is 42.8. *

2. For each of these, complete the sentence to show one way you could figure it out. Write the answer.

a. Double 31.2 is the same as double 31 plus double 0.2 * 31.2 + 31.2 = 62.4

b. Double 25.4 is the same as double 25 plus double 0.4 * 25.4 + 25.4 = 50.8

c. Double 12.3 is the same as double 12 plus double 0.3 * 12.3 + 12.3 = 24.6

d. Double 45.1 is the same as double 45 plus double 0.1 * 45.1 + 45.1 = 90.2

3. Double each of these. Write the number sentence.

a. (14.4) 14.4 + 14.4 = 28.8 b. (23.2) 23.2 + 23.2 = 46.4

c. (31.3) 31.3 + 31.3 = 62.6 d. (41.1) 41.1 + 41.1 = 82.2

4. Write some doubles that have an answer between 34 and 36.

a. 17.4 + 17.4 = 34.8 * b. 17.2 + 17.2 = 34.4 *

c. 17.6 + 17.6 = 35.2 * d. 17.8 + 17.8 = 35.6 *

Use Doubles 11

* Answers will vary. This is one example.

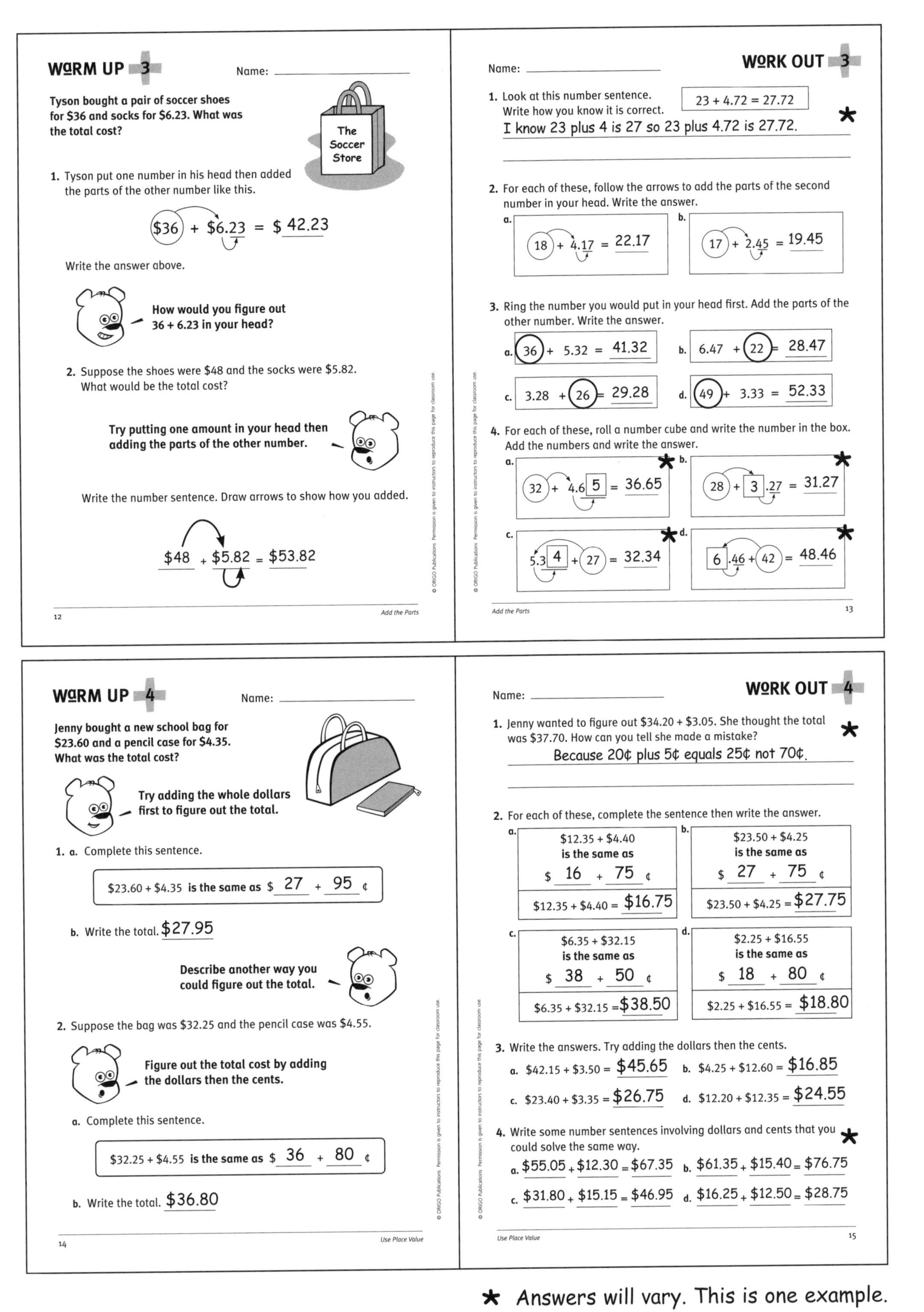

WARM UP 3

Name: ______________

Tyson bought a pair of soccer shoes for $36 and socks for $6.23. What was the total cost?

The Soccer Store

1. Tyson put one number in his head then added the parts of the other number like this.

   ($36) + $6.23 = $ 42.23

   Write the answer above.

   How would you figure out 36 + 6.23 in your head?

2. Suppose the shoes were $48 and the socks were $5.82. What would be the total cost?

   Try putting one amount in your head then adding the parts of the other number.

   Write the number sentence. Draw arrows to show how you added.

   $48 + $5.82 = $53.82

12 *Add the Parts*

WORK OUT 3

Name: ______________

1. Look at this number sentence. Write how you know it is correct. 23 + 4.72 = 27.72 *

   I know 23 plus 4 is 27 so 23 plus 4.72 is 27.72.

2. For each of these, follow the arrows to add the parts of the second number in your head. Write the answer.

   a. (18) + 4.17 = 22.17
   b. (17) + 2.45 = 19.45

3. Ring the number you would put in your head first. Add the parts of the other number. Write the answer.

   a. (36) + 5.32 = 41.32
   b. 6.47 + (22) = 28.47
   c. 3.28 + (26) = 29.28
   d. (49) + 3.33 = 52.33

4. For each of these, roll a number cube and write the number in the box. Add the numbers and write the answer.

   a. (32) + 4.6 [5] = 36.65 *
   b. (28) + [3] .27 = 31.27 *
   c. 5.3 [4] + (27) = 32.34 *
   d. [6] .46 + (42) = 48.46 *

Add the Parts 13

WARM UP 4

Name: ______________

Jenny bought a new school bag for $23.60 and a pencil case for $4.35. What was the total cost?

Try adding the whole dollars first to figure out the total.

1. a. Complete this sentence.

   $23.60 + $4.35 **is the same as** $ 27 + 95 ¢

   b. Write the total. $27.95

   Describe another way you could figure out the total.

2. Suppose the bag was $32.25 and the pencil case was $4.55.

   Figure out the total cost by adding the dollars then the cents.

   a. Complete this sentence.

   $32.25 + $4.55 **is the same as** $ 36 + 80 ¢

   b. Write the total. $36.80

14 *Use Place Value*

WORK OUT 4

Name: ______________

1. Jenny wanted to figure out $34.20 + $3.05. She thought the total was $37.70. How can you tell she made a mistake? *

   Because 20¢ plus 5¢ equals 25¢ not 70¢.

2. For each of these, complete the sentence then write the answer.

   a. $12.35 + $4.40 **is the same as** $ 16 + 75 ¢ — $12.35 + $4.40 = $16.75
   b. $23.50 + $4.25 **is the same as** $ 27 + 75 ¢ — $23.50 + $4.25 = $27.75
   c. $6.35 + $32.15 **is the same as** $ 38 + 50 ¢ — $6.35 + $32.15 = $38.50
   d. $2.25 + $16.55 **is the same as** $ 18 + 80 ¢ — $2.25 + $16.55 = $18.80

3. Write the answers. Try adding the dollars then the cents.

   a. $42.15 + $3.50 = $45.65
   b. $4.25 + $12.60 = $16.85
   c. $23.40 + $3.35 = $26.75
   d. $12.20 + $12.35 = $24.55

4. Write some number sentences involving dollars and cents that you could solve the same way. *

   a. $55.05 + $12.30 = $67.35
   b. $61.35 + $15.40 = $76.75
   c. $31.80 + $15.15 = $46.95
   d. $16.25 + $12.50 = $28.75

Use Place Value 15

* Answers will vary. This is one example.

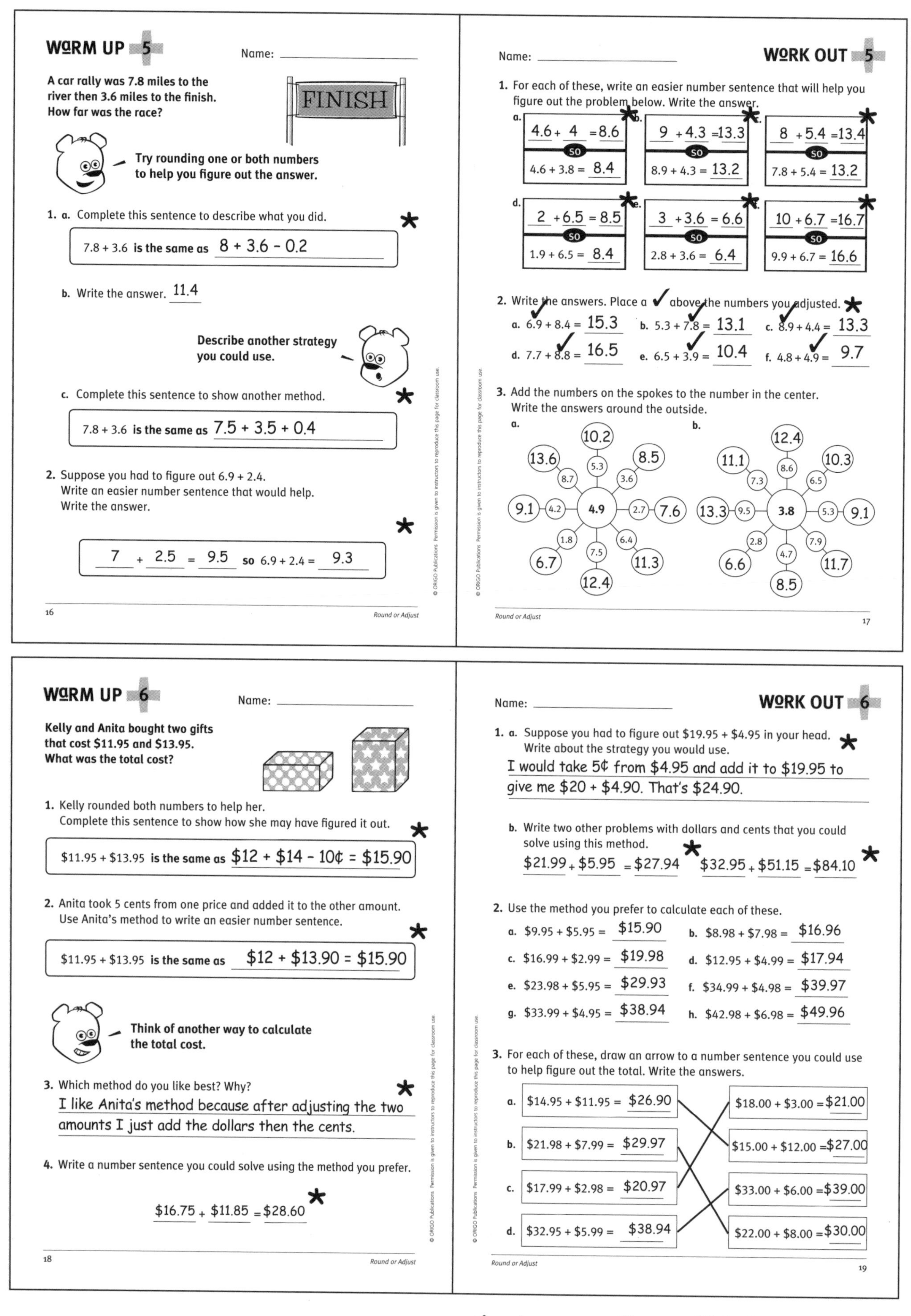

WARM UP 5

Name:

A car rally was 7.8 miles to the river then 3.6 miles to the finish. How far was the race?

Try rounding one or both numbers to help you figure out the answer.

1. a. Complete this sentence to describe what you did. *

7.8 + 3.6 **is the same as** 8 + 3.6 − 0.2

b. Write the answer. 11.4

Describe another strategy you could use.

c. Complete this sentence to show another method. *

7.8 + 3.6 **is the same as** 7.5 + 3.5 + 0.4

2. Suppose you had to figure out 6.9 + 2.4.
Write an easier number sentence that would help.
Write the answer. *

7 + 2.5 = 9.5 **so** 6.9 + 2.4 = 9.3

16 *Round or Adjust*

WORK OUT 5

Name:

1. For each of these, write an easier number sentence that will help you figure out the problem below. Write the answer.

a. 4.6 + 4 = 8.6 * **so** 4.6 + 3.8 = 8.4

b. 9 + 4.3 = 13.3 * **so** 8.9 + 4.3 = 13.2

c. 8 + 5.4 = 13.4 * **so** 7.8 + 5.4 = 13.2

d. 2 + 6.5 = 8.5 * **so** 1.9 + 6.5 = 8.4

e. 3 + 3.6 = 6.6 * **so** 2.8 + 3.6 = 6.4

f. 10 + 6.7 = 16.7 * **so** 9.9 + 6.7 = 16.6

2. Write the answers. Place a ✓ above the numbers you adjusted. *

a. 6.9 + 8.4 = 15.3 b. 5.3 + 7.8 = 13.1 c. 8.9 + 4.4 = 13.3

d. 7.7 + 8.8 = 16.5 e. 6.5 + 3.9 = 10.4 f. 4.8 + 4.9 = 9.7

3. Add the numbers on the spokes to the number in the center.
Write the answers around the outside.

a. Center 4.9; spokes 5.3, 3.6, 2.7, 6.4, 7.5, 1.8, 4.2, 8.7; answers 10.2, 8.5, 7.6, 11.3, 12.4, 6.7, 9.1, 13.6

b. Center 3.8; spokes 8.6, 6.5, 5.3, 7.9, 4.7, 2.8, 9.5, 7.3; answers 12.4, 10.3, 9.1, 11.7, 8.5, 6.6, 13.3, 11.1

Round or Adjust 17

WARM UP 6

Name:

Kelly and Anita bought two gifts that cost $11.95 and $13.95. What was the total cost?

1. Kelly rounded both numbers to help her.
Complete this sentence to show how she may have figured it out. *

$11.95 + $13.95 **is the same as** $12 + $14 − 10¢ = $15.90

2. Anita took 5 cents from one price and added it to the other amount.
Use Anita's method to write an easier number sentence. *

$11.95 + $13.95 **is the same as** $12 + $13.90 = $15.90

Think of another way to calculate the total cost.

3. Which method do you like best? Why? *

I like Anita's method because after adjusting the two amounts I just add the dollars then the cents.

4. Write a number sentence you could solve using the method you prefer. *

$16.75 + $11.85 = $28.60

18 *Round or Adjust*

WORK OUT 6

Name:

1. a. Suppose you had to figure out $19.95 + $4.95 in your head. *
Write about the strategy you would use.

I would take 5¢ from $4.95 and add it to $19.95 to give me $20 + $4.90. That's $24.90.

b. Write two other problems with dollars and cents that you could solve using this method.

$21.99 + $5.95 = $27.94 * $32.95 + $51.15 = $84.10 *

2. Use the method you prefer to calculate each of these.

a. $9.95 + $5.95 = $15.90 b. $8.98 + $7.98 = $16.96

c. $16.99 + $2.99 = $19.98 d. $12.95 + $4.99 = $17.94

e. $23.98 + $5.95 = $29.93 f. $34.99 + $4.98 = $39.97

g. $33.99 + $4.95 = $38.94 h. $42.98 + $6.98 = $49.96

3. For each of these, draw an arrow to a number sentence you could use to help figure out the total. Write the answers.

a. $14.95 + $11.95 = $26.90 → $15.00 + $12.00 = $27.00

b. $21.98 + $7.99 = $29.97 → $22.00 + $8.00 = $30.00

c. $17.99 + $2.98 = $20.97 → $18.00 + $3.00 = $21.00

d. $32.95 + $5.99 = $38.94 → $33.00 + $6.00 = $39.00

Round or Adjust 19

* Answers will vary. This is one example.

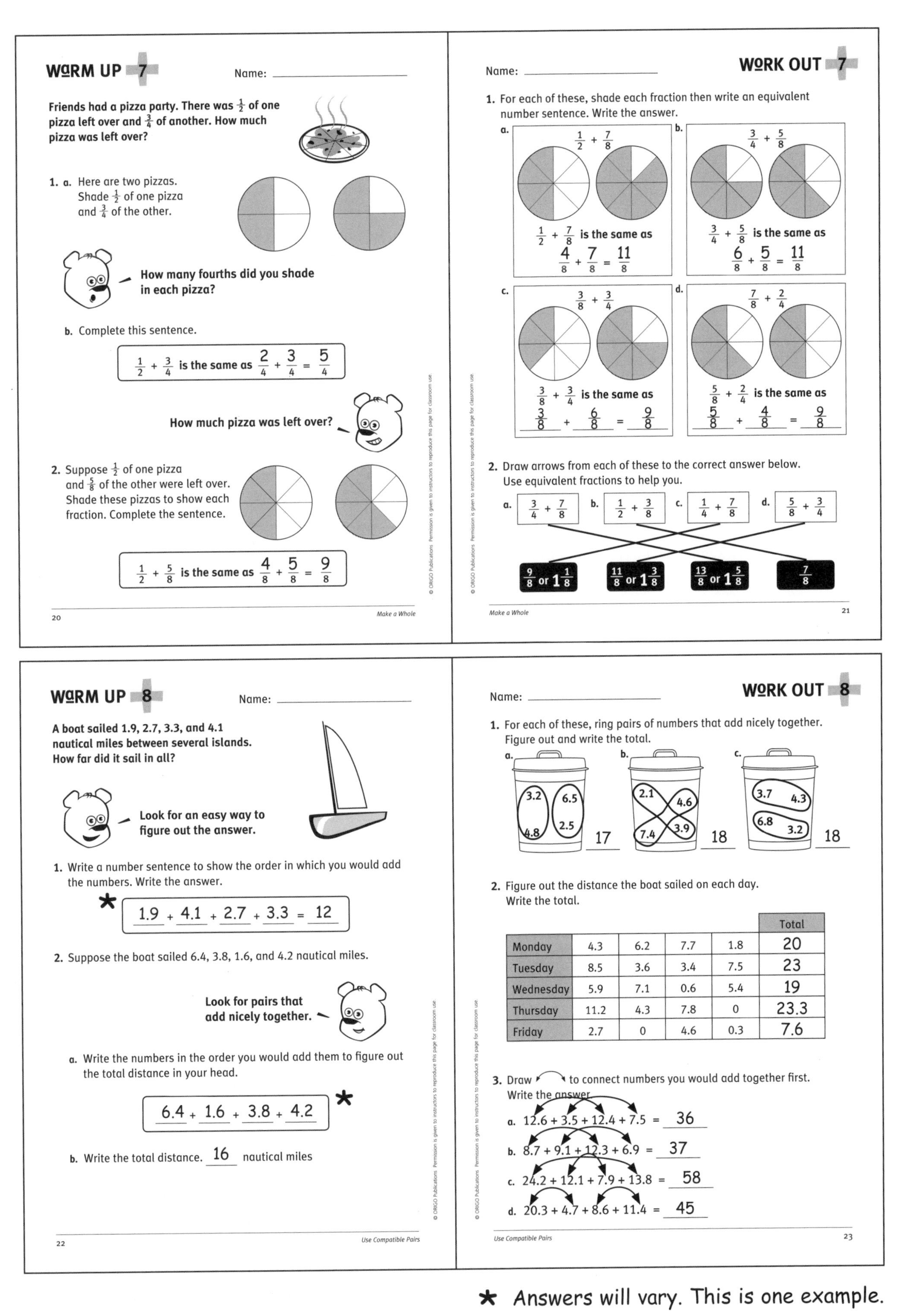

WARM UP 7

Name: ____________

Friends had a pizza party. There was $\frac{1}{2}$ of one pizza left over and $\frac{3}{4}$ of another. How much pizza was left over?

1. a. Here are two pizzas. Shade $\frac{1}{2}$ of one pizza and $\frac{3}{4}$ of the other.

How many fourths did you shade in each pizza?

b. Complete this sentence.

$\frac{1}{2} + \frac{3}{4}$ is the same as $\frac{2}{4} + \frac{3}{4} = \frac{5}{4}$

How much pizza was left over?

2. Suppose $\frac{1}{2}$ of one pizza and $\frac{5}{8}$ of the other were left over. Shade these pizzas to show each fraction. Complete the sentence.

$\frac{1}{2} + \frac{5}{8}$ is the same as $\frac{4}{8} + \frac{5}{8} = \frac{9}{8}$

20 *Make a Whole*

WORK OUT 7

Name: ____________

1. For each of these, shade each fraction then write an equivalent number sentence. Write the answer.

a. $\frac{1}{2} + \frac{7}{8}$ is the same as $\frac{4}{8} + \frac{7}{8} = \frac{11}{8}$

b. $\frac{3}{4} + \frac{5}{8}$ is the same as $\frac{6}{8} + \frac{5}{8} = \frac{11}{8}$

c. $\frac{3}{8} + \frac{3}{4}$ is the same as $\frac{3}{8} + \frac{6}{8} = \frac{9}{8}$

d. $\frac{7}{8} + \frac{2}{4}$ is the same as $\frac{5}{8} + \frac{4}{8} = \frac{9}{8}$

2. Draw arrows from each of these to the correct answer below. Use equivalent fractions to help you.

a. $\frac{3}{4} + \frac{7}{8}$ b. $\frac{1}{2} + \frac{3}{8}$ c. $\frac{1}{4} + \frac{7}{8}$ d. $\frac{5}{8} + \frac{3}{4}$

$\frac{9}{8}$ or $1\frac{1}{8}$ $\frac{11}{8}$ or $1\frac{3}{8}$ $\frac{13}{8}$ or $1\frac{5}{8}$ $\frac{7}{8}$

© ORIGO Publications Permission is given to instructors to reproduce this page for classroom use.

Make a Whole 21

WARM UP 8

Name: ____________

A boat sailed 1.9, 2.7, 3.3, and 4.1 nautical miles between several islands. How far did it sail in all?

Look for an easy way to figure out the answer.

1. Write a number sentence to show the order in which you would add the numbers. Write the answer.

★ 1.9 + 4.1 + 2.7 + 3.3 = 12

2. Suppose the boat sailed 6.4, 3.8, 1.6, and 4.2 nautical miles.

Look for pairs that add nicely together.

a. Write the numbers in the order you would add them to figure out the total distance in your head.

6.4 + 1.6 + 3.8 + 4.2 ★

b. Write the total distance. 16 nautical miles

© ORIGO Publications Permission is given to instructors to reproduce this page for classroom use.

22 *Use Compatible Pairs*

WORK OUT 8

Name: ____________

1. For each of these, ring pairs of numbers that add nicely together. Figure out and write the total.

a. 3.2, 6.5, 4.8, 2.5 — 17

b. 2.1, 4.6, 7.4, 3.9 — 18

c. 3.7, 4.3, 6.8, 3.2 — 18

2. Figure out the distance the boat sailed on each day. Write the total.

					Total
Monday	4.3	6.2	7.7	1.8	20
Tuesday	8.5	3.6	3.4	7.5	23
Wednesday	5.9	7.1	0.6	5.4	19
Thursday	11.2	4.3	7.8	0	23.3
Friday	2.7	0	4.6	0.3	7.6

3. Draw ⌒ to connect numbers you would add together first. Write the answer.

a. 12.6 + 3.5 + 12.4 + 7.5 = 36

b. 8.7 + 9.1 + 12.3 + 6.9 = 37

c. 24.2 + 12.1 + 7.9 + 13.8 = 58

d. 20.3 + 4.7 + 8.6 + 11.4 = 45

© ORIGO Publications Permission is given to instructors to reproduce this page for classroom use.

Use Compatible Pairs 23

★ Answers will vary. This is one example.

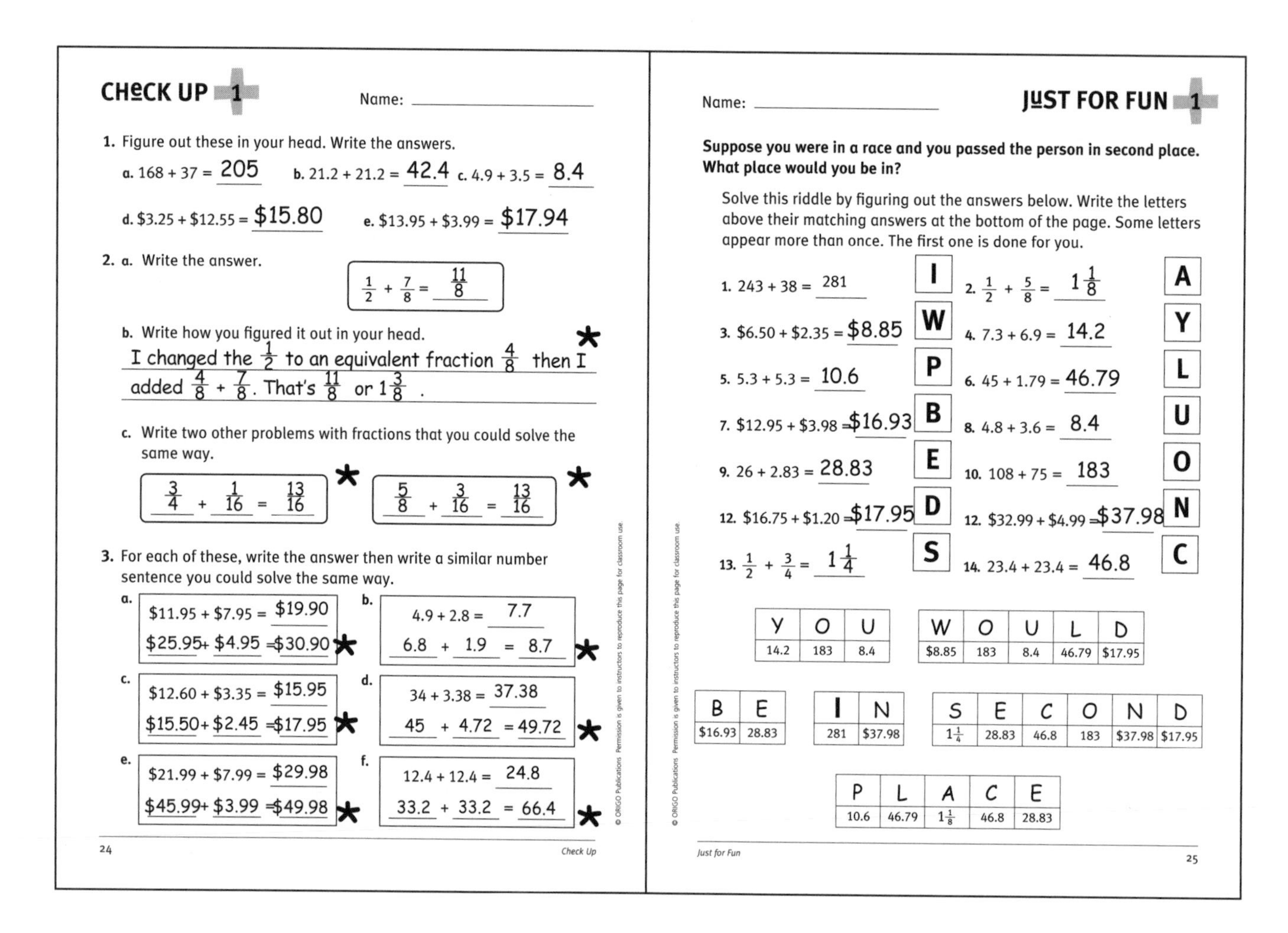

CHECK UP 1

Name: ____________________

1. Figure out these in your head. Write the answers.

   a. 168 + 37 = 205 b. 21.2 + 21.2 = 42.4 c. 4.9 + 3.5 = 8.4

   d. \$3.25 + \$12.55 = \$15.80 e. \$13.95 + \$3.99 = \$17.94

2. a. Write the answer.

   $\frac{1}{2} + \frac{7}{8} = \frac{11}{8}$

   b. Write how you figured it out in your head. *

   I changed the $\frac{1}{2}$ to an equivalent fraction $\frac{4}{8}$ then I added $\frac{4}{8} + \frac{7}{8}$. That's $\frac{11}{8}$ or $1\frac{3}{8}$.

   c. Write two other problems with fractions that you could solve the same way.

   $\frac{3}{4} + \frac{1}{16} = \frac{13}{16}$ * $\frac{5}{8} + \frac{3}{16} = \frac{13}{16}$ *

3. For each of these, write the answer then write a similar number sentence you could solve the same way.

   a. \$11.95 + \$7.95 = \$19.90
   \$25.95 + \$4.95 = \$30.90 *

   b. 4.9 + 2.8 = 7.7
   6.8 + 1.9 = 8.7 *

   c. \$12.60 + \$3.35 = \$15.95
   \$15.50 + \$2.45 = \$17.95 *

   d. 34 + 3.38 = 37.38
   45 + 4.72 = 49.72 *

   e. \$21.99 + \$7.99 = \$29.98
   \$45.99 + \$3.99 = \$49.98 *

   f. 12.4 + 12.4 = 24.8
   33.2 + 33.2 = 66.4 *

JUST FOR FUN 1

Name: ____________________

Suppose you were in a race and you passed the person in second place. What place would you be in?

Solve this riddle by figuring out the answers below. Write the letters above their matching answers at the bottom of the page. Some letters appear more than once. The first one is done for you.

1. 243 + 38 = 281 **I**
2. $\frac{1}{2} + \frac{5}{8} = 1\frac{1}{8}$ **A**
3. \$6.50 + \$2.35 = \$8.85 **W**
4. 7.3 + 6.9 = 14.2 **Y**
5. 5.3 + 5.3 = 10.6 **P**
6. 45 + 1.79 = 46.79 **L**
7. \$12.95 + \$3.98 = \$16.93 **B**
8. 4.8 + 3.6 = 8.4 **U**
9. 26 + 2.83 = 28.83 **E**
10. 108 + 75 = 183 **O**
12. \$16.75 + \$1.20 = \$17.95 **D**
12. \$32.99 + \$4.99 = \$37.98 **N**
13. $\frac{1}{2} + \frac{3}{4} = 1\frac{1}{4}$ **S**
14. 23.4 + 23.4 = 46.8 **C**

Y	O	U
14.2	183	8.4

W	O	U	L	D
\$8.85	183	8.4	46.79	\$17.95

B	E
\$16.93	28.83

I	N
281	\$37.98

S	E	C	O	N	D
$1\frac{1}{4}$	28.83	46.8	183	\$37.98	\$17.95

P	L	A	C	E
10.6	46.79	$1\frac{1}{8}$	46.8	28.83

* Answers will vary. This is one example.

SUBTRACTION STRATEGIES

Count back

293 – 44 is 293 – 10 – 10 – 10 – 10 – 4

Count on

\$140 – \$65 = \$75 because \$65 + \$5 + \$70 = \$140

\$3.80 – \$2.75 = \$1.05 because \$2.75 + 5¢ + \$1 = \$3.80

$4 - 2\frac{3}{4} = 1\frac{1}{4}$ *because* $2\frac{3}{4} + \frac{1}{4} + 1 = 4$

Subtract the parts

\$27 – \$4.50 is the same as \$27 – \$4 – 50¢

Use place value

69.4 – 32.2 is the same as (60 – 30) + (9 – 2) + (0.4 – 0.2)

or (69 – 32) + (0.4 – 0.2)

or (0.4 – 0.2) + (9 – 2) + (60 – 30)

Round or adjust

\$158 – \$49 is the same as \$158 – \$50 + \$1

\$10 – \$2.49 is the same as \$10 – \$2.50 + 1¢

WaRM UP 9

Name: ______________________

Elliot had 293 trading cards in his collection. He gave 44 cards away. How many cards does he have left?

1. a. Draw jumps on this number line to count back 44.

220 230 240 250 260 270 280 290 300 310 320

b. Write a number sentence to show what you did.

293 – ______________________

Think of another way you could count back 44.

c. Draw jumps again to show another way.

2. a. Suppose Elliot had 284 cards and gave away 37. Show how you would count back 37.

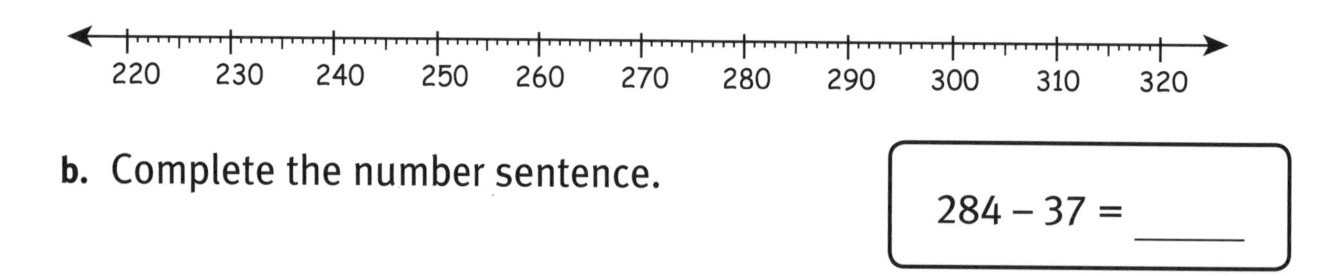

b. Complete the number sentence.

284 – 37 = ______

Name: ______________________

WORK OUT 9

1. Figure out each of these in your head. Write the answer.
Draw jumps to show how you did it.

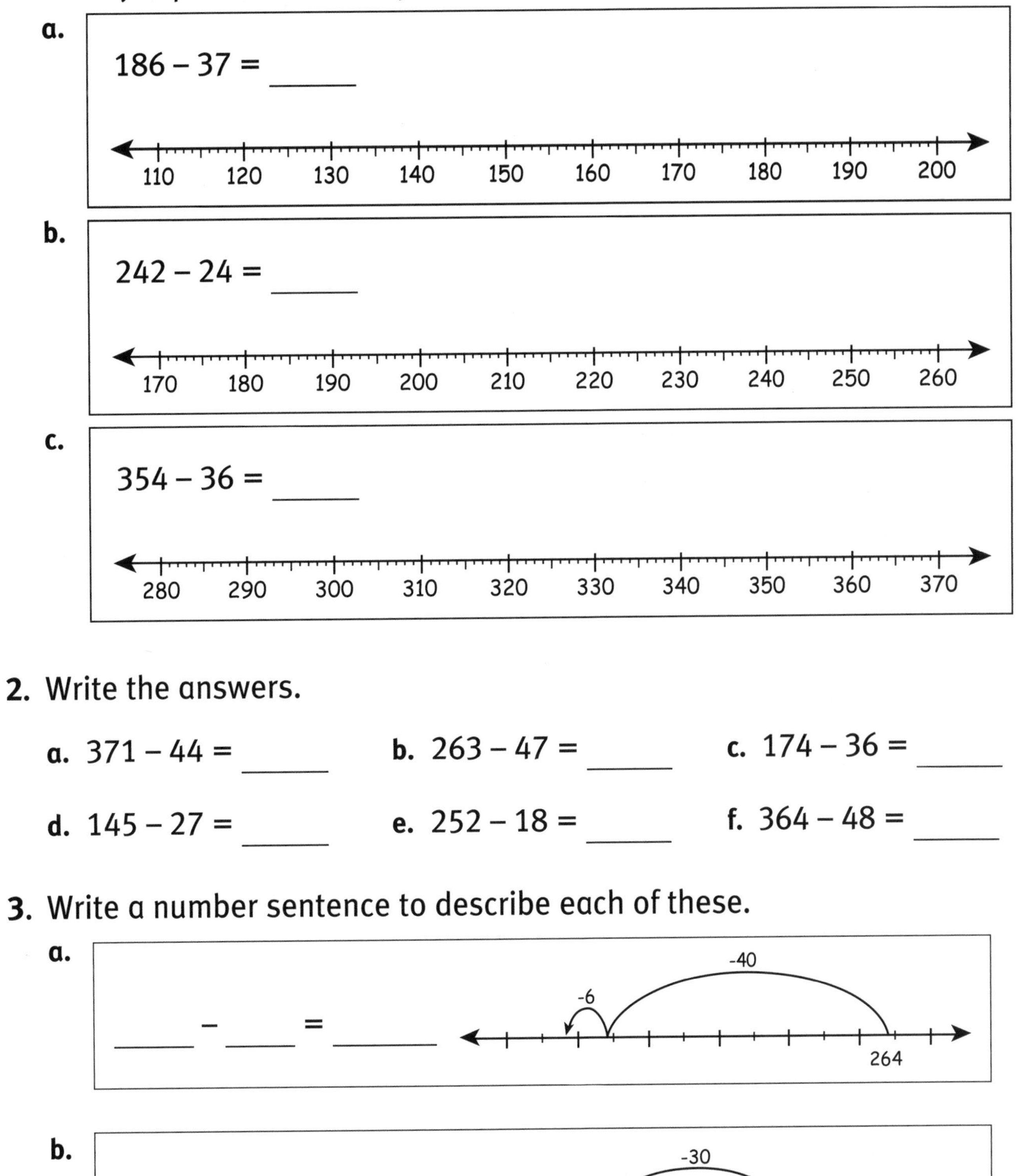

a. 186 – 37 = ______

b. 242 – 24 = ______

c. 354 – 36 = ______

2. Write the answers.

a. 371 – 44 = ______ **b.** 263 – 47 = ______ **c.** 174 – 36 = ______

d. 145 – 27 = ______ **e.** 252 – 18 = ______ **f.** 364 – 48 = ______

3. Write a number sentence to describe each of these.

a. ______ – ______ = ______

b. ______ – ______ = ______

WARM UP 10

Name: ______________________

Amy has a coupon for $140. If she buys a CD player for $65, how much will she have left to spend?

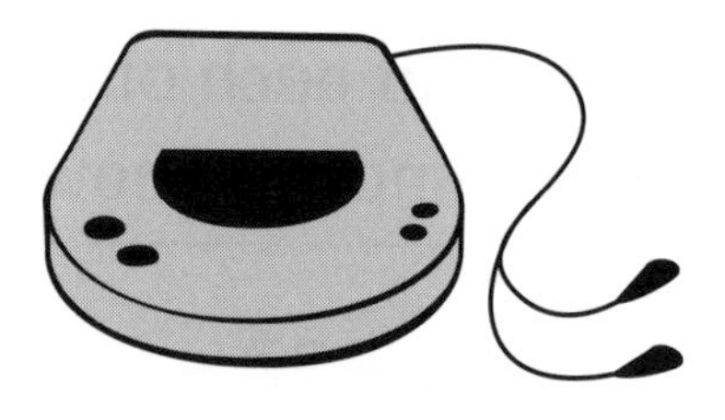

1. a. Draw jumps to show how you could start at 65 and count on to 140.

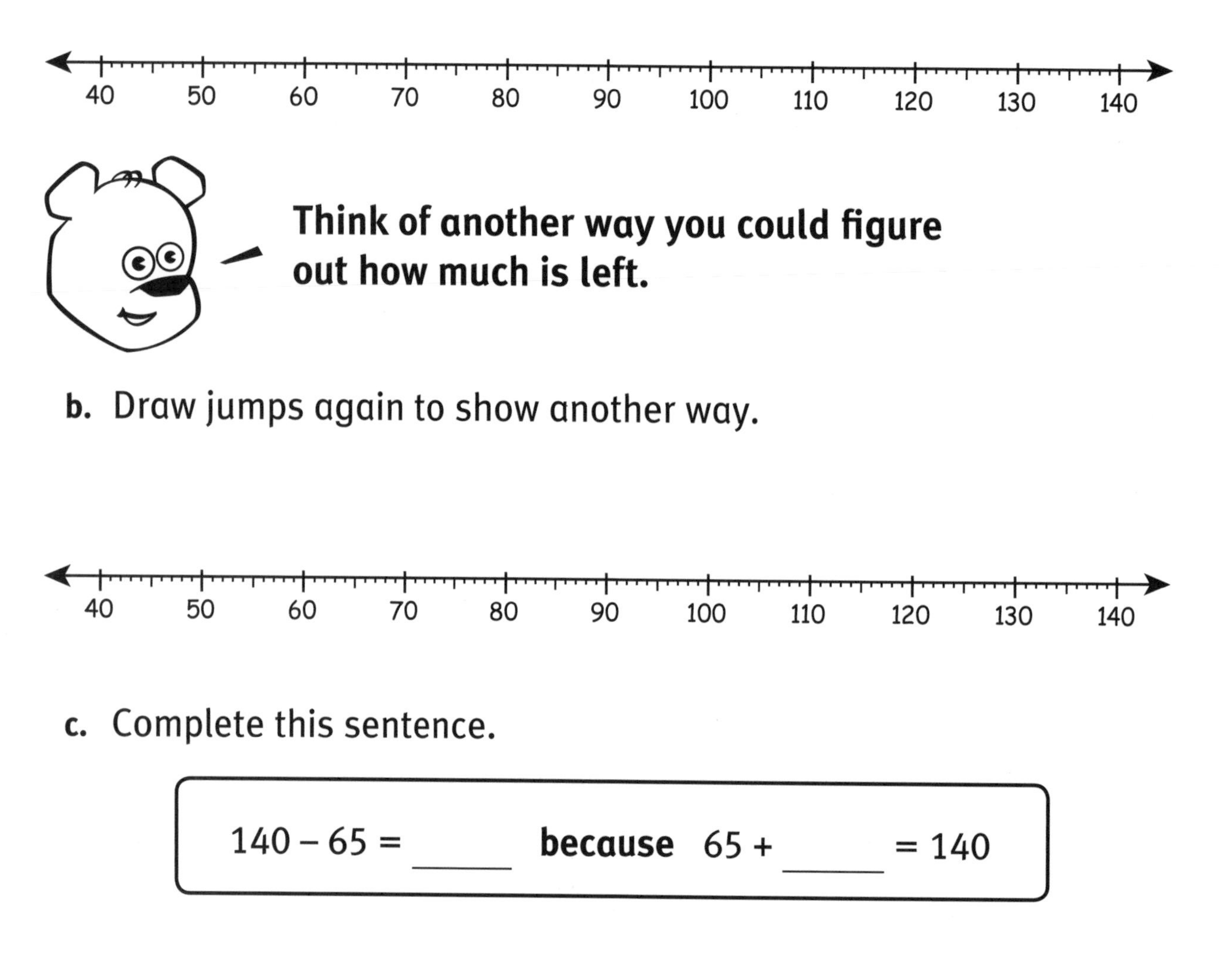

Think of another way you could figure out how much is left.

b. Draw jumps again to show another way.

40 50 60 70 80 90 100 110 120 130 140

c. Complete this sentence.

140 – 65 = ______ **because** 65 + ______ = 140

2. Suppose the CD player was $78. Write a number sentence to show how much she would have left to spend.

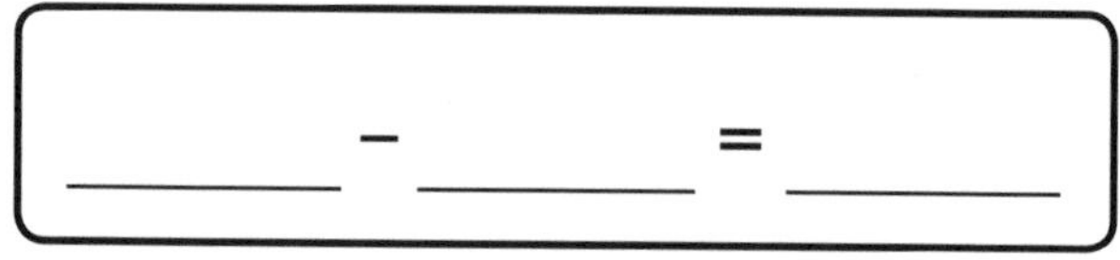

Name: ____________________

1. Figure out each of these in your head. Write the answer.
Draw jumps to show how you did it.

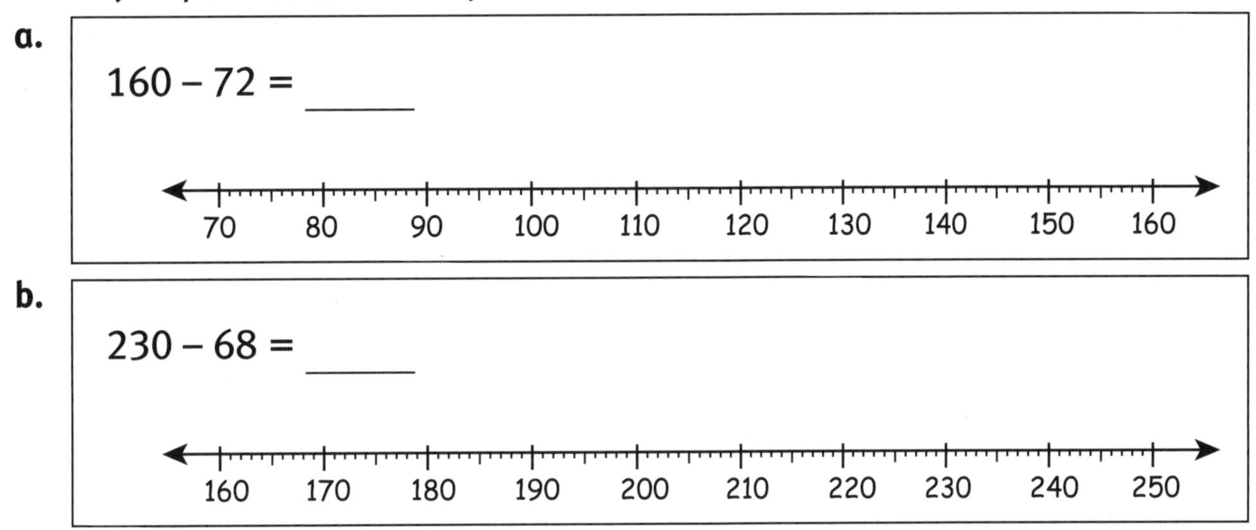

2. Look at these sale items.

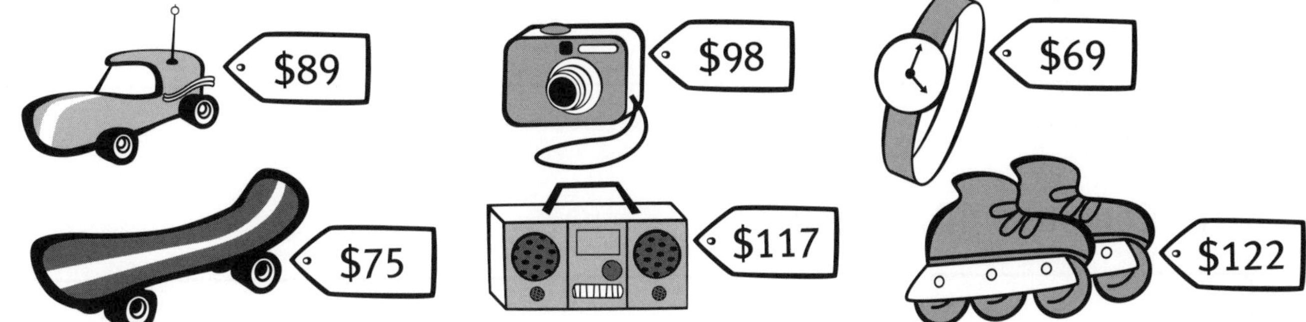

For each coupon, write how much would be left to spend
after buying each item.

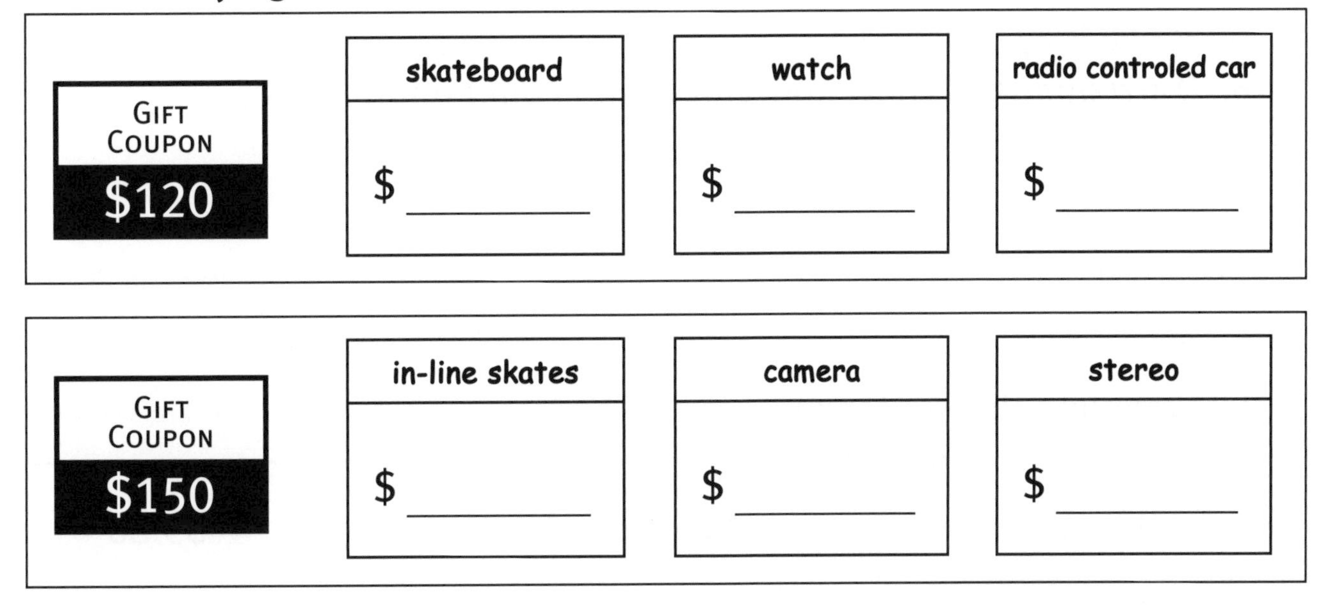

WaRM UP 11

Name: ______________________

Joey has $2.75. He needs $3.80 to buy a sandwich. How much more money does he need?

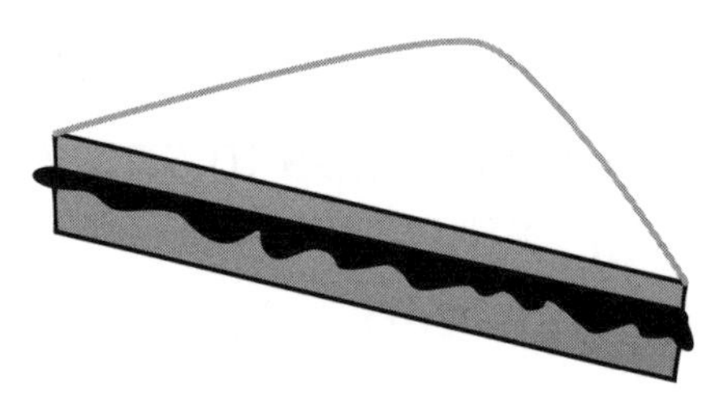

1. This is one way you could count on to $3.80.

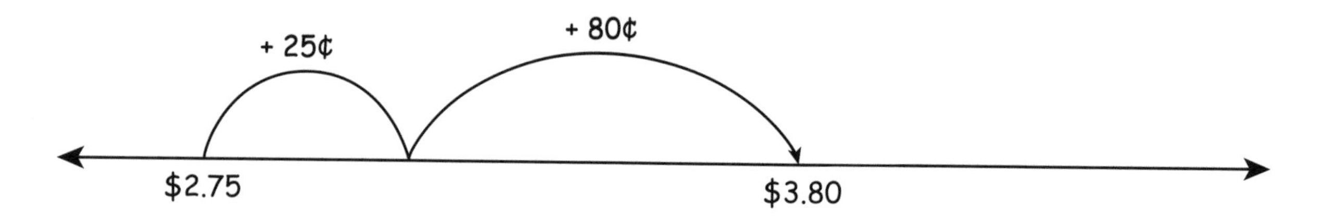

a. Draw jumps like those above to show another way to count on.

$2.75

b. Complete this sentence.

$3.80 – $2.75 = ________ **because** $2.75 + ________ = $3.80

2. Suppose Joey only had $2.35.

a. Draw jumps to show how you would count on to $3.80.

$2.35

b. Complete this sentence.

$3.80 – $2.35 = ________ **because** $2.35 + ________ = $3.80

Name: ____________________

1. Figure out each of these in your head. Write the answer. Draw jumps to show how you did it.

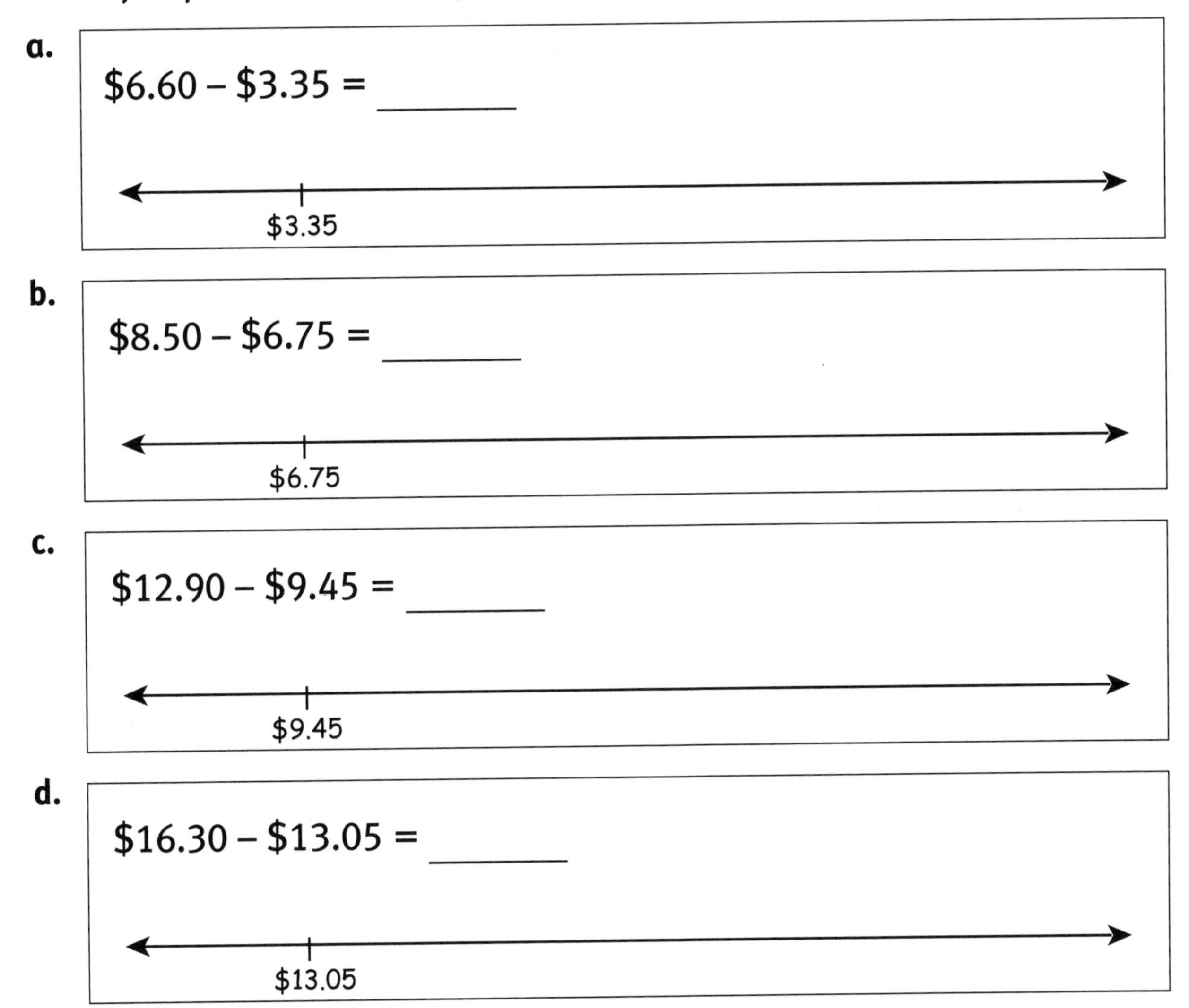

a. $6.60 – $3.35 = ______

b. $8.50 – $6.75 = ______

c. $12.90 – $9.45 = ______

d. $16.30 – $13.05 = ______

2. Write a number sentence to describe each of these.

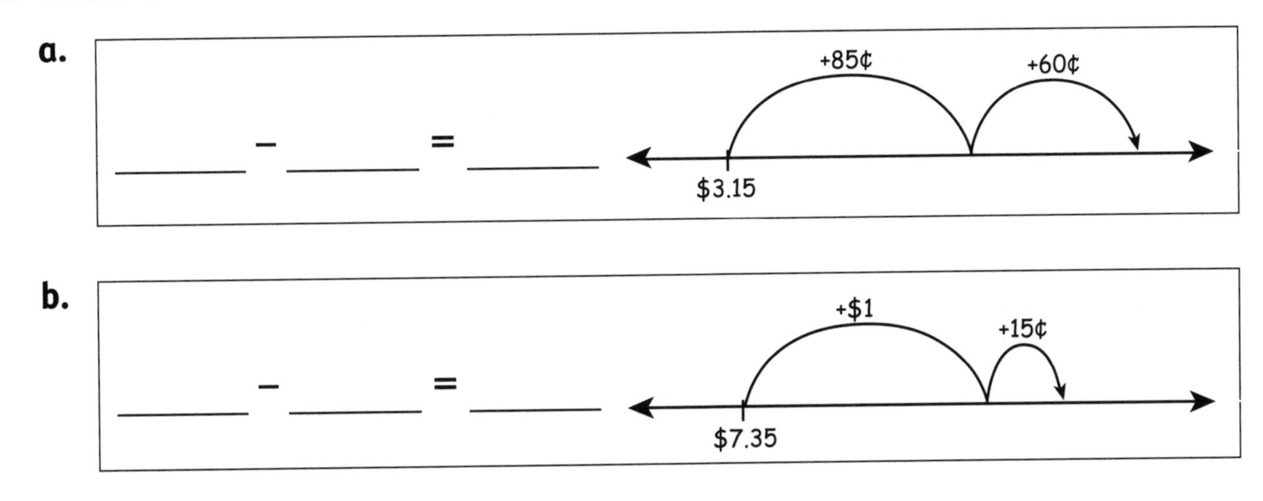

a. ______ – ______ = ______

b. ______ – ______ = ______

WARM UP 12

Name: ____________________

**Chris had $27. He spent $4.50 on lunch.
How much money does he have left?**

1. a. This picture shows $27.
   Cross out $4.50.

   b. How much is left? ________

What part did you cross out first - the dollars or the cents?

2. a. How much money

   is in this box? ________

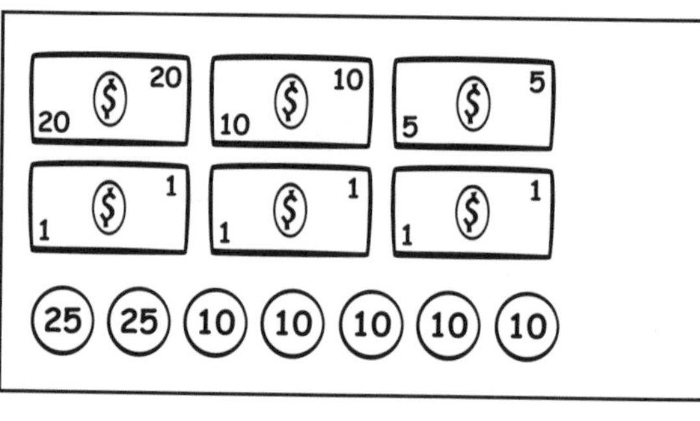

   b. Cross out $7.35. Write a number sentence to show what you did.

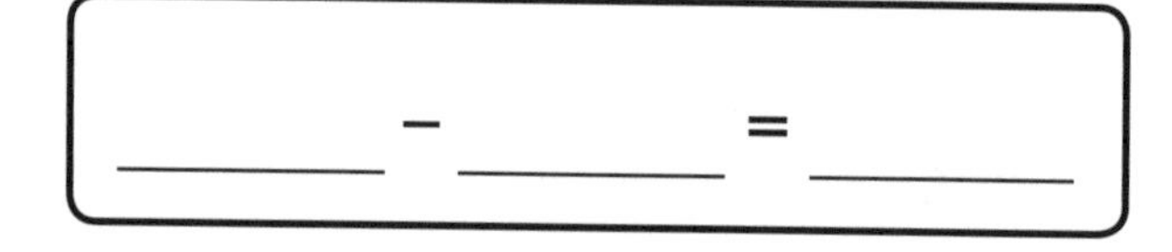

3. Cross out $12.15 from this box.
   Write a number sentence to show what you did.

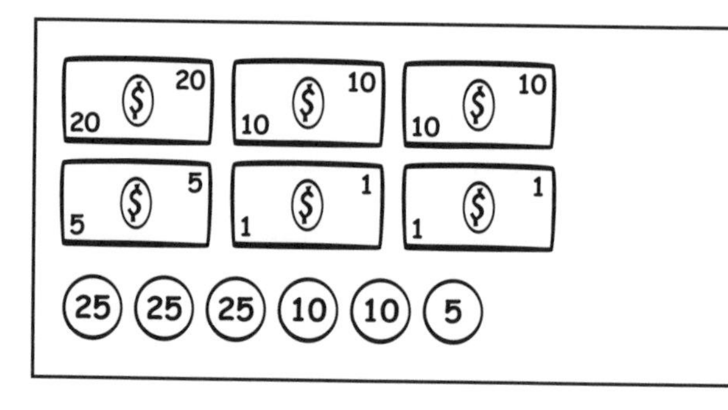

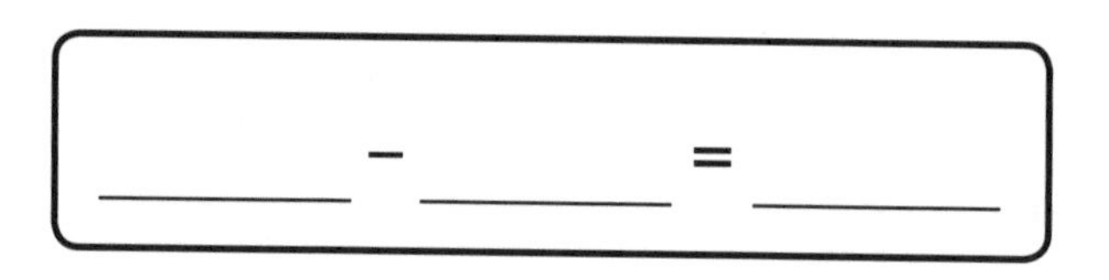

Name: ____________________

WORK OUT 12

1. For each of these, cross out the amount shown. Complete the number sentence.

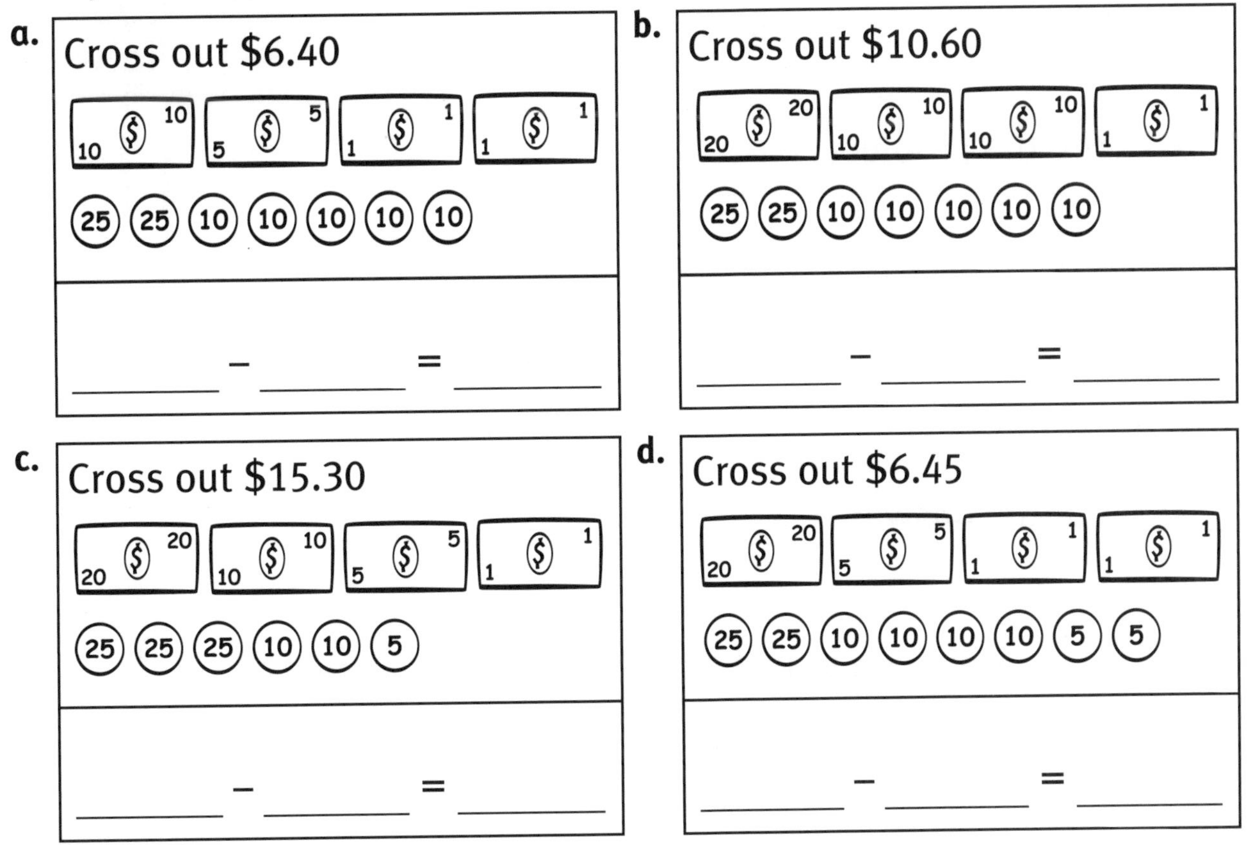

2. Spend the amount shown. Write the new amount. Try subtracting the dollars first.

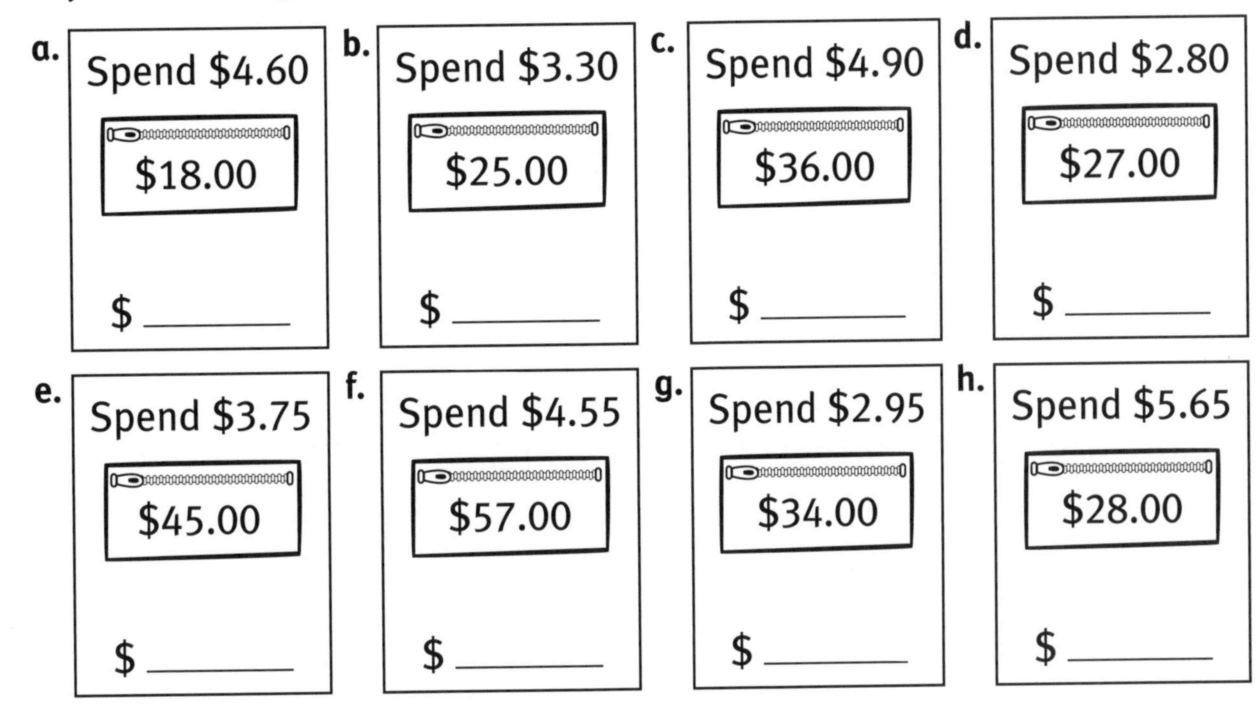

WaRM UP 13

Name: ______________

It is 69.4 miles to the beach by the scenic road and only 32.2 miles by the highway. How much shorter is the highway?

1. a. Write the problem. _______ – _______

What is an easy way to figure out problems such as this?

b. Write the answer. _______

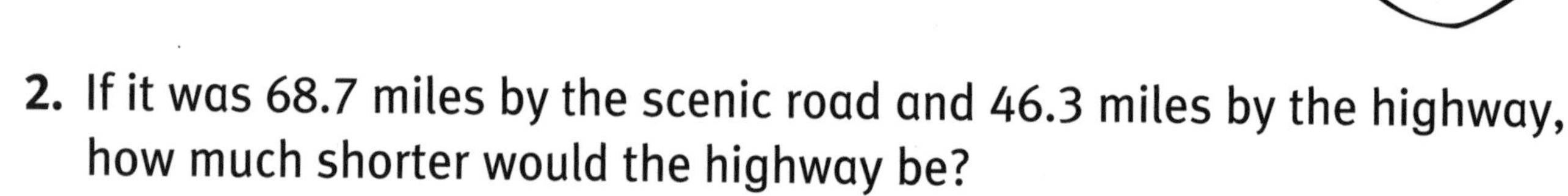

2. If it was 68.7 miles by the scenic road and 46.3 miles by the highway, how much shorter would the highway be?

Look for an easy way to figure out the answer.

Write the number sentence.

_______ – _______ = _______

3. Write two subtraction sentences involving tenths that you could solve the same way.

a.

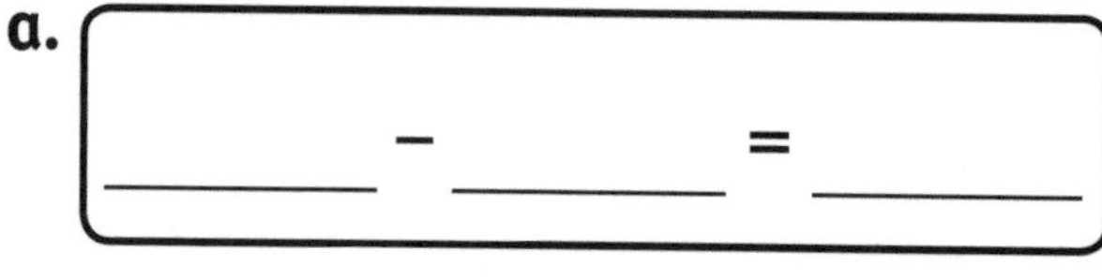

b.

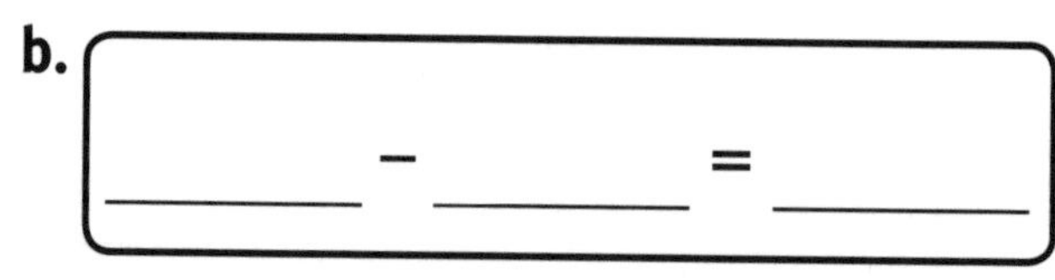

Name: ____________________

WORK OUT 13

1. a. Suppose you had to figure out 47.6 – 23.4 in your head. Write how you would do it.

____

____

b. Describe another way you could do it.

____

____

c. Write three subtraction problems you could solve using either of these two methods.

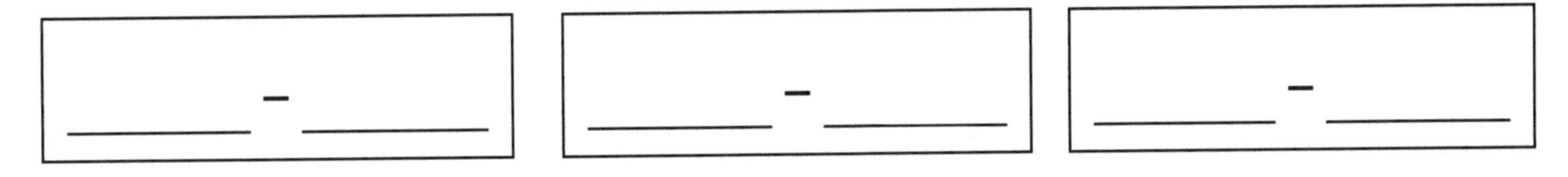

2. Use the method you prefer to calculate each of these.

a. 75.9 – 22.5 = ________ **b.** 86.8 – 63.4 = ________

c. 34.7 – 13.4 = ________ **d.** 59.3 – 47.1 = ________

e. 256.4 – 132.3 = ________ **f.** 563.7 – 241.6 = ________

3. Write the missing number in each pyramid.
The sum of two numbers must equal the number above.

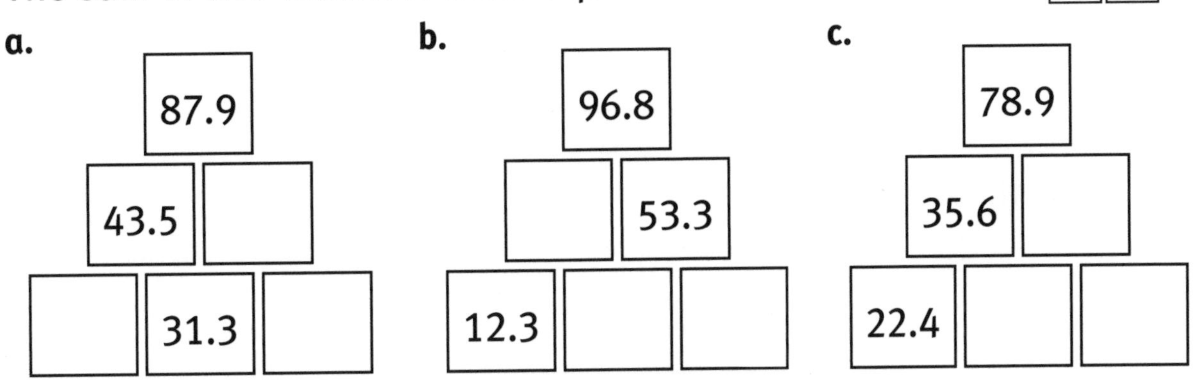

W<u>a</u>RM UP 14

Name: ____________________

Ryan had $158 in savings. He bought new basketball shoes for $49. How much money does he have left?

How could you use 158 – 50 to help you figure out 158 – 49?

1. Describe your method. Complete this sentence to help you.

158 – 50 = _____ **so** 158 – 49 = _____

2. What is an easy way to figure out 128 – 59 in your head?
Complete this sentence to help you.

128 – 60 = _____ **so** 128 – 59 = _____

Describe another strategy you could use.

3. a. Suppose you had to figure out 228 – 39.
Write an easier number sentence you know would help.

_____ – _____ = _____

b. Write the answer.

228 – 39 = _____

Name: ______________________

WORK OUT 14

1. For each of these, write an easier number sentence that will help you figure out the answer. Write the answer.

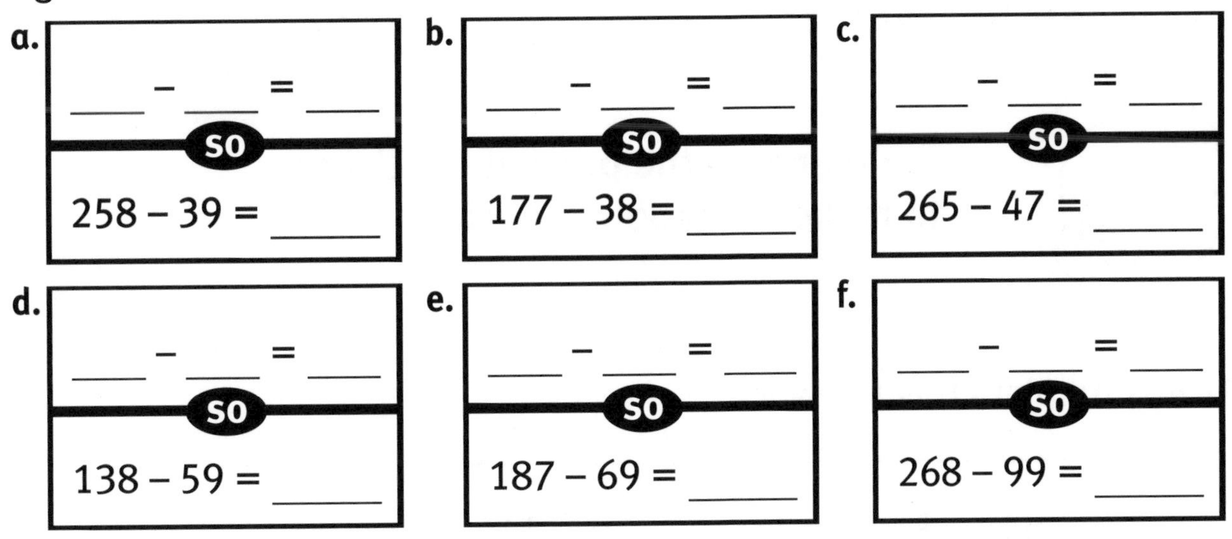

2. Write the answers. Place a ✔ above the numbers you adjusted.

a. 167 - 49 = ______ **b.** 348 – 69 = ______ **c.** 236 – 78 = ______

3. Take the numbers on the spokes away from the number in the center. Write the answers around the outside.

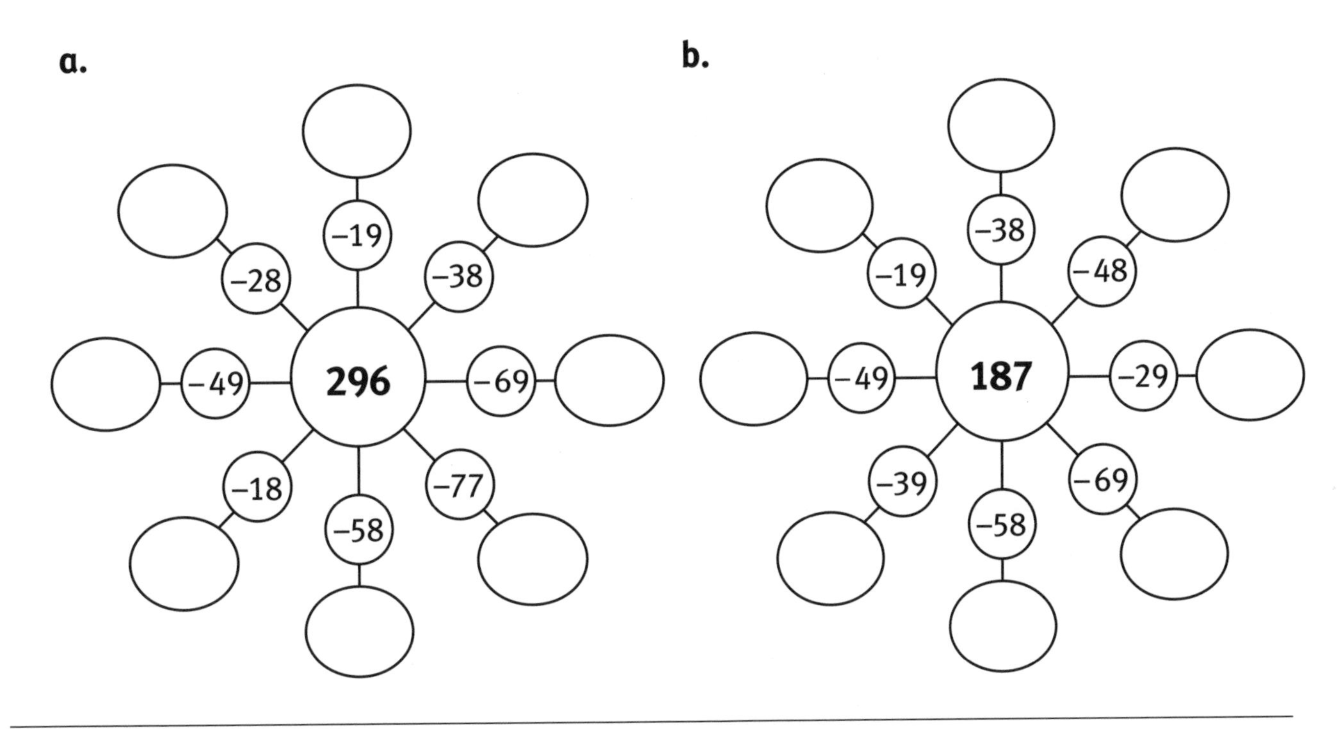

WaRM UP 15

Name: ______________________

Mick had $10 and spent $2.49 to buy a train ticket. How much money does he have left?

Try rounding the price of the ticket to make the problem easier.

1. a. Complete this sentence.

$10 – $2.50 = ________ **so** $10 – $2.49 = ________

Describe a different way to calculate the answer.

b. Complete this sentence to show another method.

$10 – $2.49 **is the same as** ______________________

2. Suppose you had to figure out $5 – $2.98. Write an easier number sentence that would help. Write the answer.

________ – ________ = ________ **so** $5 – $2.98 = ________

Name: ______________________

1. For each of these, write an easier number sentence that will help you figure out the answer. Write the answer.

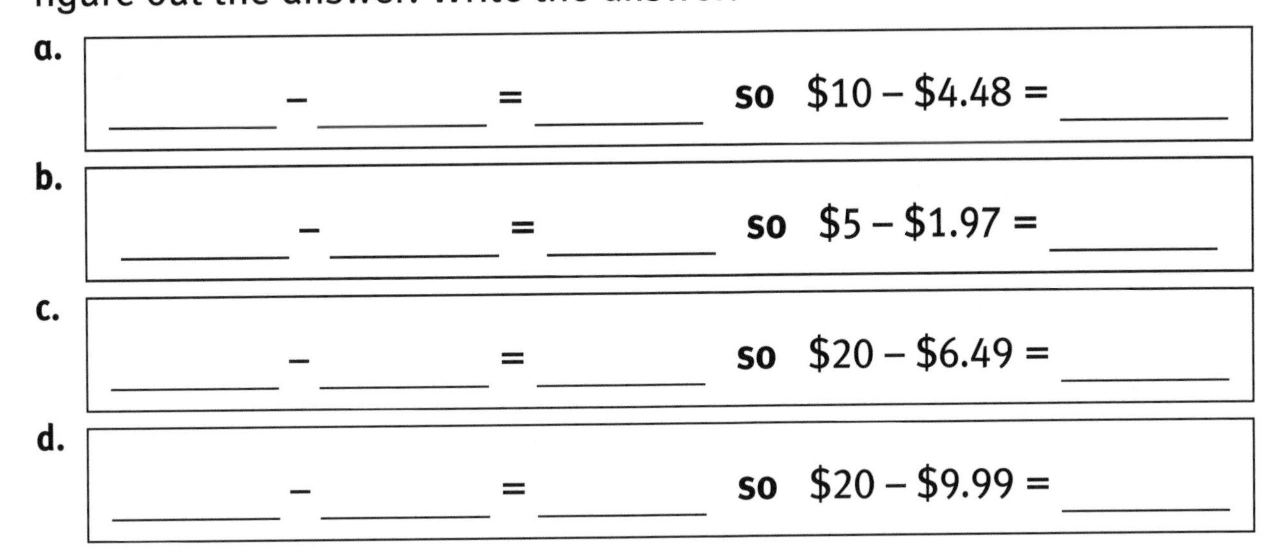

a. ______ – ______ = ______ so \$10 – \$4.48 = ______

b. ______ – ______ = ______ so \$5 – \$1.97 = ______

c. ______ – ______ = ______ so \$20 – \$6.49 = ______

d. ______ – ______ = ______ so \$20 – \$9.99 = ______

2. Look at this grocery sale.

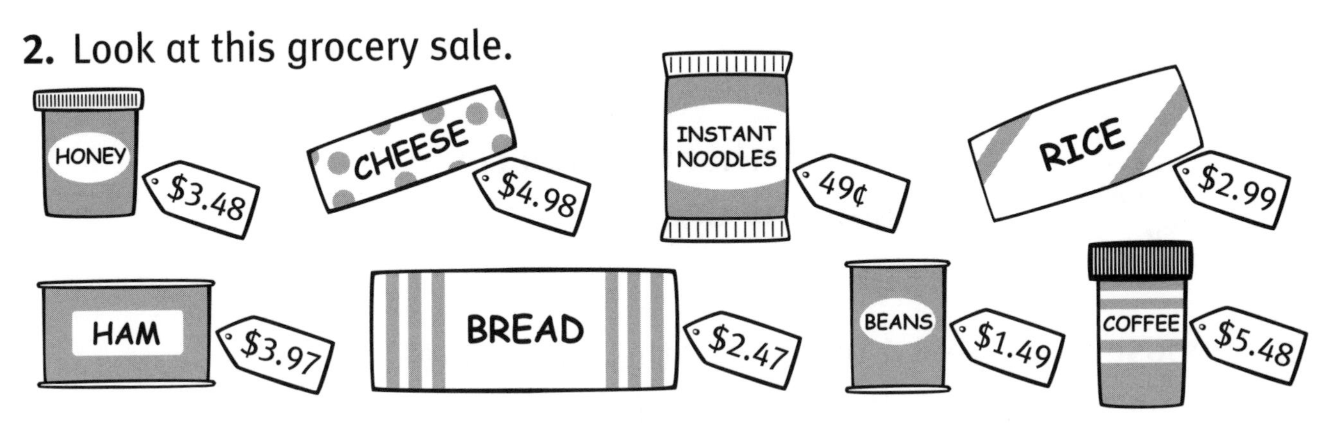

For each amount, write how much will be left to spend after buying the grocery item.

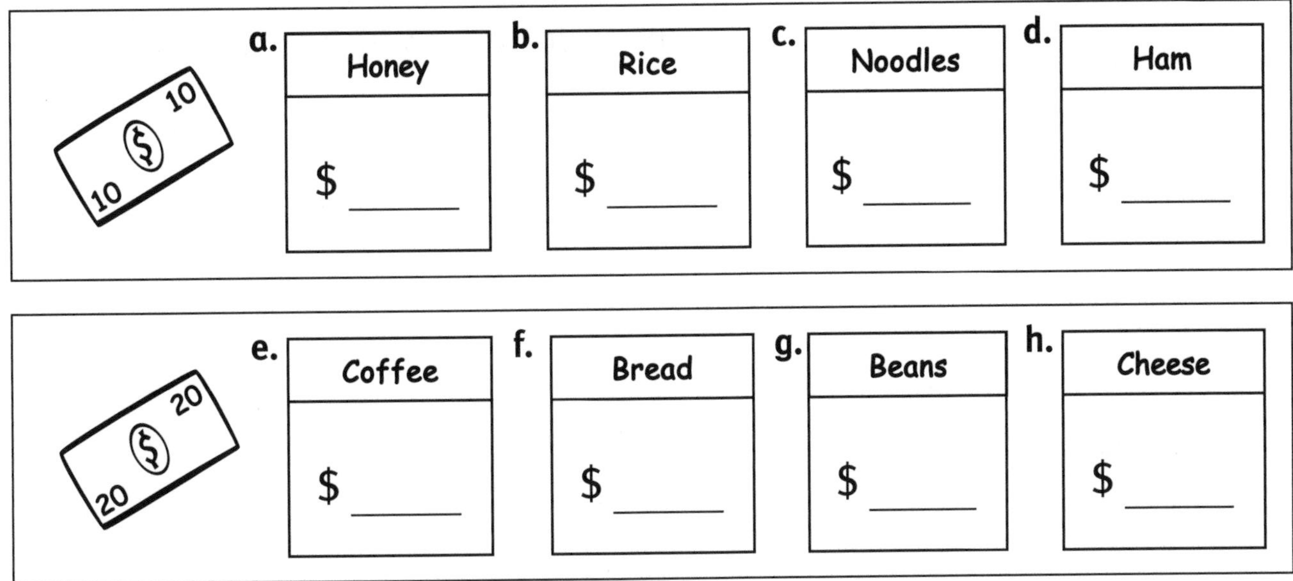

	a. Honey	b. Rice	c. Noodles	d. Ham
\$10	\$ ______	\$ ______	\$ ______	\$ ______

	e. Coffee	f. Bread	g. Beans	h. Cheese
\$20	\$ ______	\$ ______	\$ ______	\$ ______

WaRM UP 16

Name: ______________________

Some friends bought 3 pizzas. They ate 2 pizzas and $\frac{3}{8}$ of the last one. How much pizza was left?

1. a. Cross out 2 pizzas then $\frac{3}{8}$ of a pizza to show how to subtract the parts.

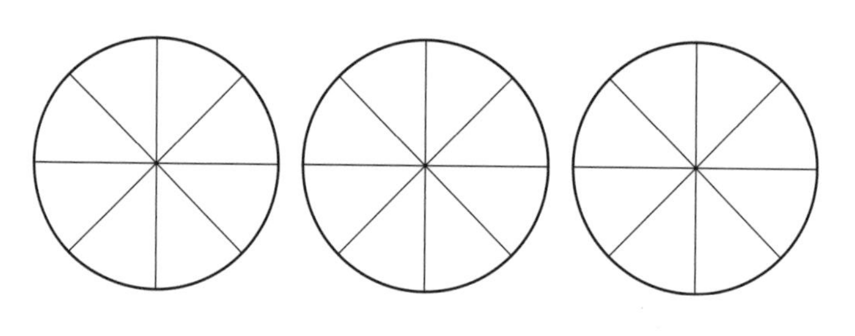

b. Write the number sentence.

_____ – _____ = _____

What is another way you could calculate the answer?

2. Suppose there were 4 pizzas and the friends ate $2\frac{3}{4}$. How much pizza would be left over?

Complete the sentence.

$4 - 2\frac{3}{4} =$ _______

Which method do you prefer? Why?

Name: ______________________

WORK OUT 16

1. Calculate $5 - 3\frac{3}{4}$ in your head. Describe the strategy you used.

____

____

2. For each of these, use your method to figure out how much is left. Complete the number sentence.

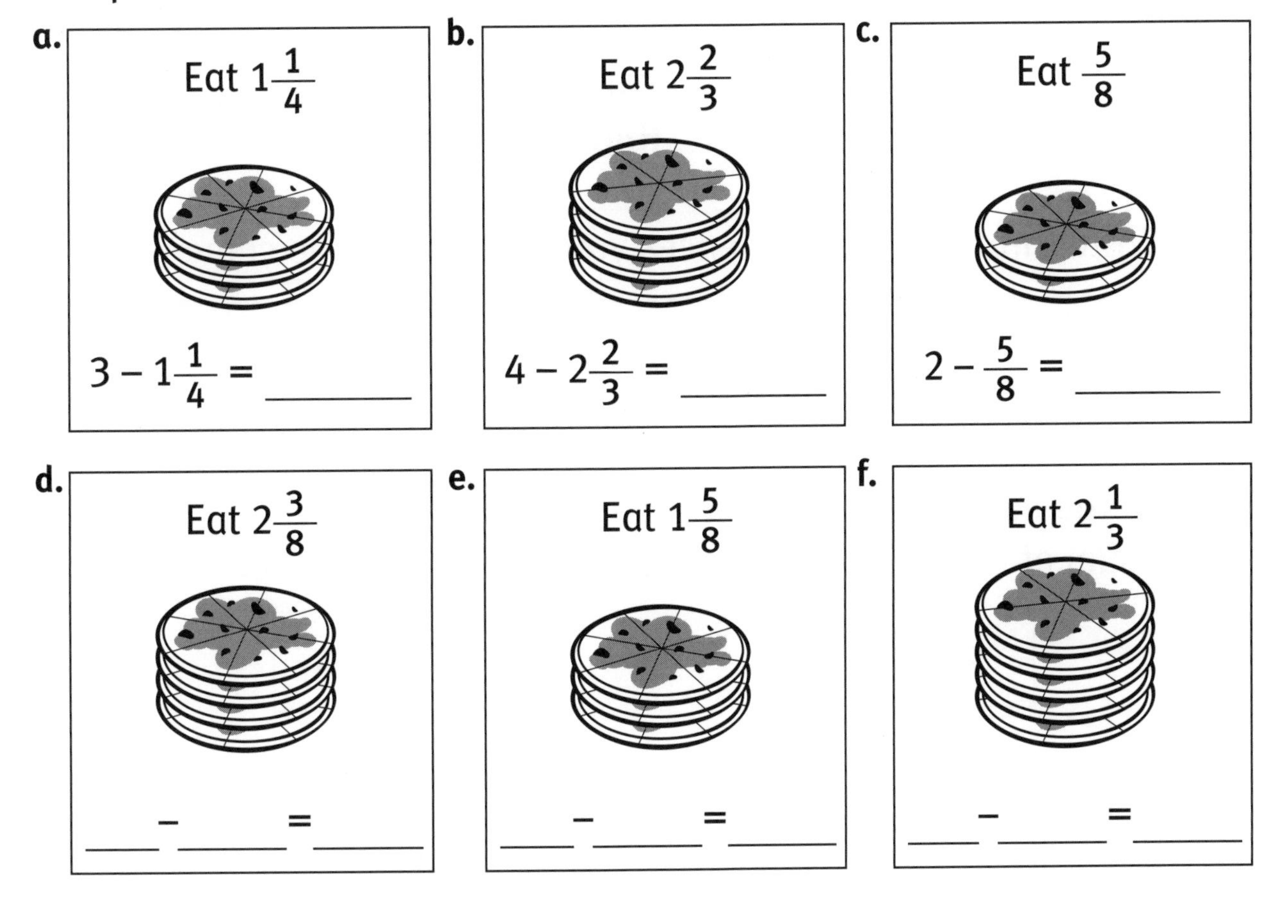

3. Calculate each of these.

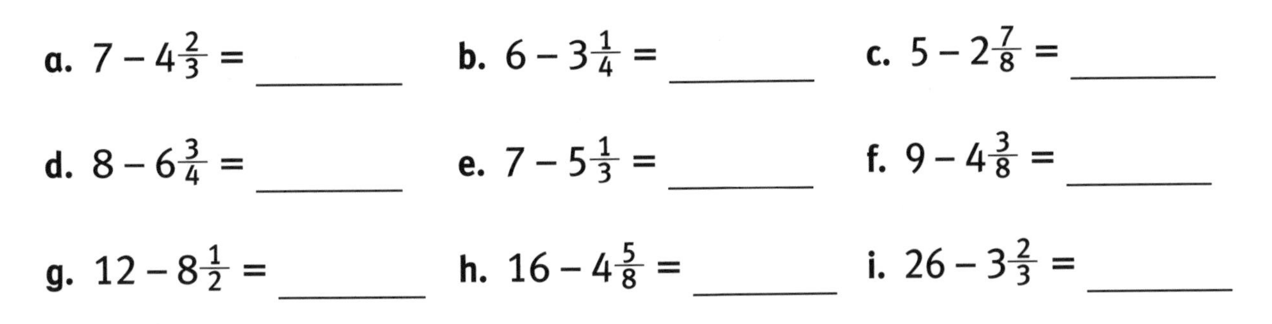

a. $7 - 4\frac{2}{3} =$ ______ **b.** $6 - 3\frac{1}{4} =$ ______ **c.** $5 - 2\frac{7}{8} =$ ______

d. $8 - 6\frac{3}{4} =$ ______ **e.** $7 - 5\frac{1}{3} =$ ______ **f.** $9 - 4\frac{3}{8} =$ ______

g. $12 - 8\frac{1}{2} =$ ______ **h.** $16 - 4\frac{5}{8} =$ ______ **i.** $26 - 3\frac{2}{3} =$ ______

CHECK UP 2

Name: ____________________

1. Figure out these in your head. Write the answers.

a. $46.7 - 23.4 =$ ______ **b.** $243 - 26 =$ ______ **c.** $220 - 68 =$ ______

d. $4 - 2\frac{1}{4} =$ ______ **e.** $177 - 38 =$ ______ **f.** $7 - 5\frac{1}{3} =$ ______

g. $\$18 - \$4.80 =$ ________ **h.** $\$5.70 - \$2.35 =$ ________

2. a. Write the answer.

$168 - 39 =$ ______

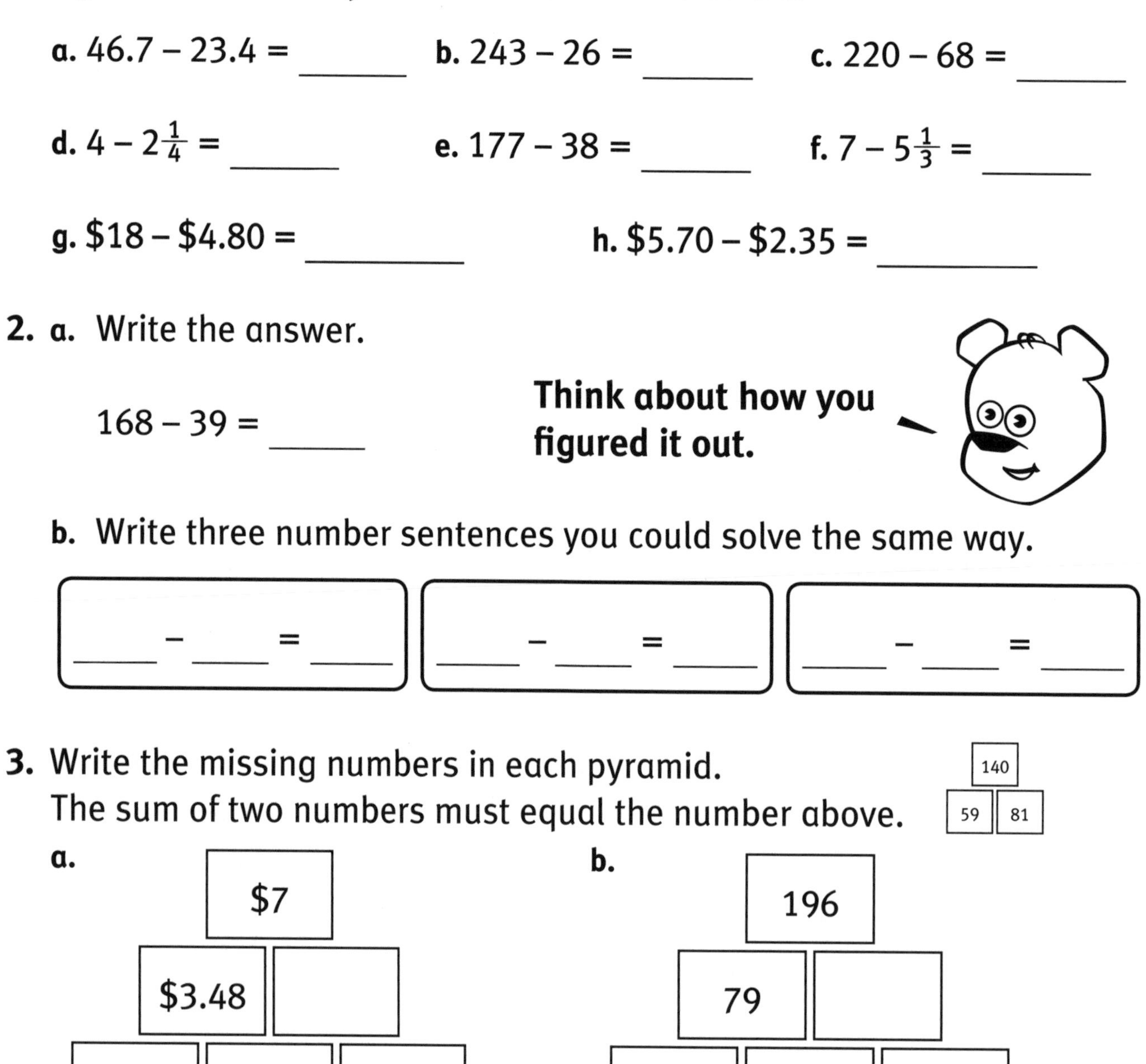

Think about how you figured it out.

b. Write three number sentences you could solve the same way.

____ − ____ = ____ ____ − ____ = ____ ____ − ____ = ____

3. Write the missing numbers in each pyramid.
The sum of two numbers must equal the number above.

(Example: 140 above 59 and 81)

a.
- Top: $7
- Middle: $3.48, ____
- Bottom: $1.28, ____, ____

b.
- Top: 196
- Middle: 79, ____
- Bottom: ____, 34, ____

c.
- Top: 88.9
- Middle: ____, 23.3
- Bottom: ____, ____, 11.1

d.
- Top: 127
- Middle: ____, 59
- Bottom: ____, 29, ____

Name: ______________________

JUST FOR FUN 2

There were two tourists standing outside the White House.
One of them was the father of the other one's son.
How could this be?

Solve this riddle by figuring out the answers below. Write the letters above their matching answers at the bottom of the page. Some letters appear more than once.

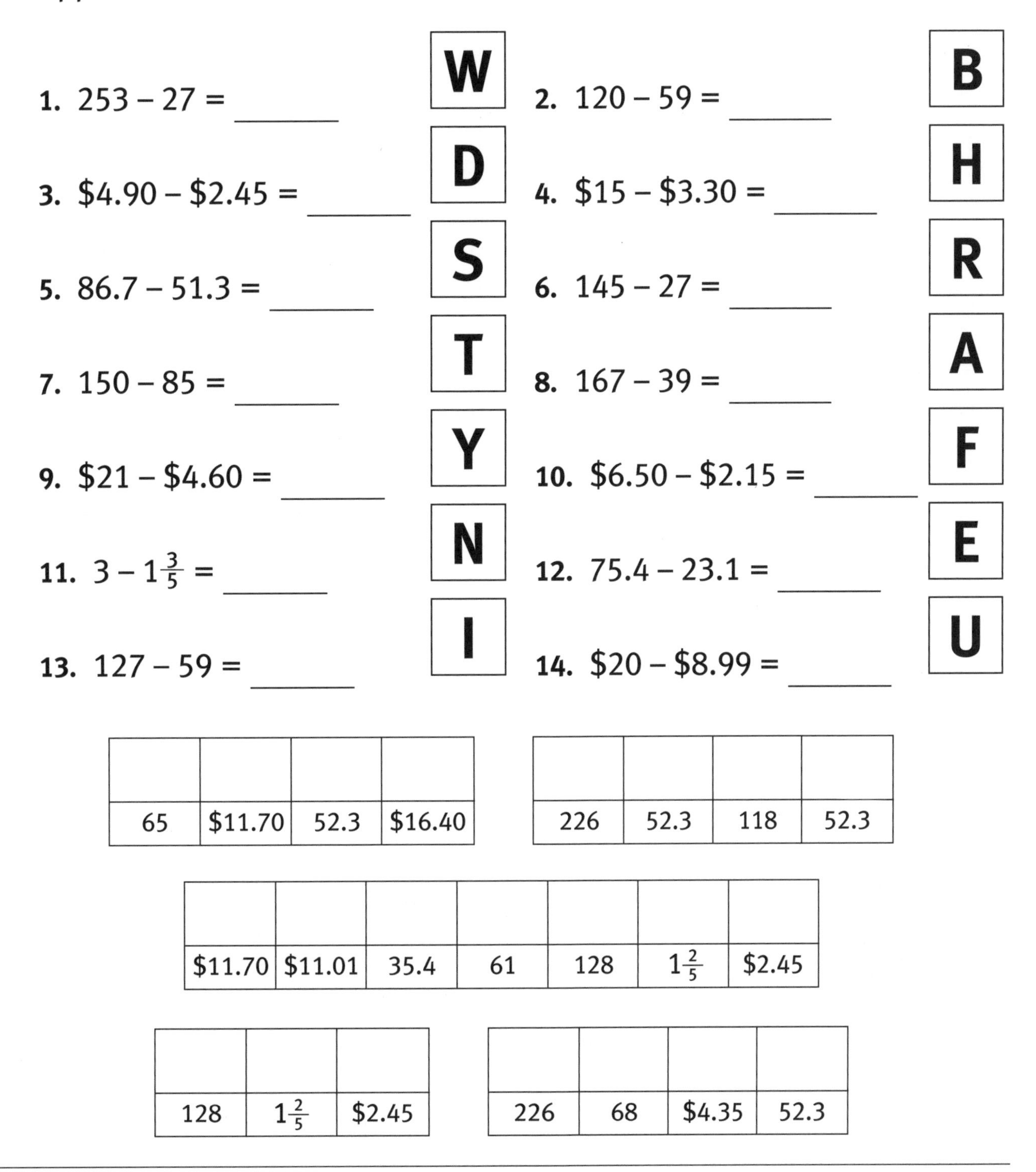

1. $253 - 27 =$ ______ **W**
2. $120 - 59 =$ ______ **B**
3. $4.90 – $2.45 = ______ **D**
4. $15 – $3.30 = ______ **H**
5. $86.7 - 51.3 =$ ______ **S**
6. $145 - 27 =$ ______ **R**
7. $150 - 85 =$ ______ **T**
8. $167 - 39 =$ ______ **A**
9. $21 – $4.60 = ______ **Y**
10. $6.50 – $2.15 = ______ **F**
11. $3 - 1\frac{3}{5} =$ ______ **N**
12. $75.4 - 23.1 =$ ______ **E**
13. $127 - 59 =$ ______ **I**
14. $20 – $8.99 = ______ **U**

65	$11.70	52.3	$16.40

226	52.3	118	52.3

$11.70	$11.01	35.4	61	128	$1\frac{2}{5}$	$2.45

128	$1\frac{2}{5}$	$2.45

226	68	$4.35	52.3

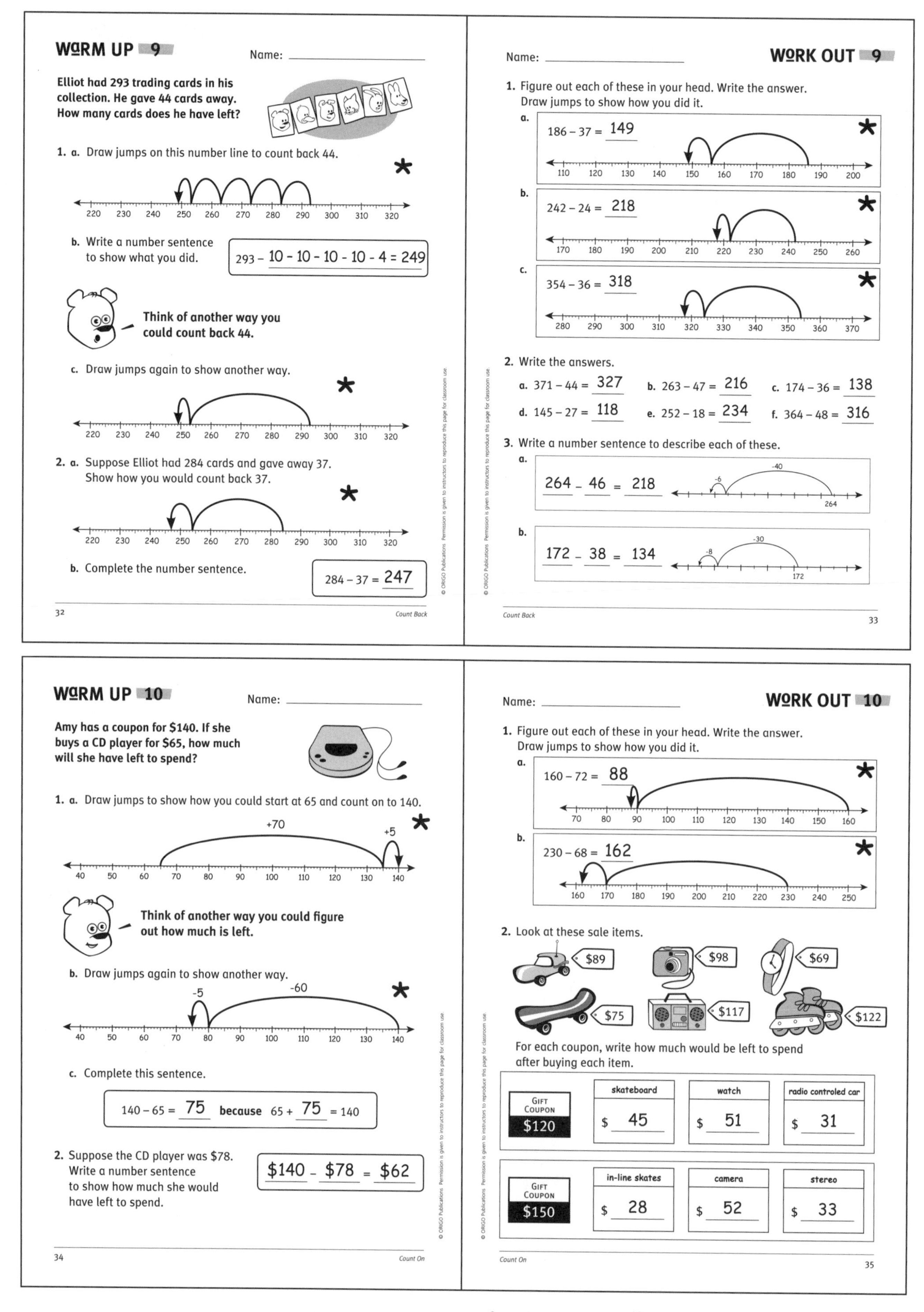
WARM UP 9

Name: ____________

Elliot had 293 trading cards in his collection. He gave 44 cards away. How many cards does he have left?

1. a. Draw jumps on this number line to count back 44. *

220 230 240 250 260 270 280 290 300 310 320

b. Write a number sentence to show what you did.

293 – 10 - 10 - 10 - 10 - 4 = 249

Think of another way you could count back 44.

c. Draw jumps again to show another way. *

220 230 240 250 260 270 280 290 300 310 320

2. a. Suppose Elliot had 284 cards and gave away 37. Show how you would count back 37. *

220 230 240 250 260 270 280 290 300 310 320

b. Complete the number sentence.

284 – 37 = 247

32 Count Back

© ORIGO Publications Permission is given to instructors to reproduce this page for classroom use.

WORK OUT 9

Name: ____________

1. Figure out each of these in your head. Write the answer. Draw jumps to show how you did it.

a. 186 – 37 = 149 *

110 120 130 140 150 160 170 180 190 200

b. 242 – 24 = 218 *

170 180 190 200 210 220 230 240 250 260

c. 354 – 36 = 318 *

280 290 300 310 320 330 340 350 360 370

2. Write the answers.

a. 371 – 44 = 327 b. 263 – 47 = 216 c. 174 – 36 = 138

d. 145 – 27 = 118 e. 252 – 18 = 234 f. 364 – 48 = 316

3. Write a number sentence to describe each of these.

a. 264 – 46 = 218 (-6, -40, 264)

b. 172 – 38 = 134 (-8, -30, 172)

© ORIGO Publications Permission is given to instructors to reproduce this page for classroom use.

Count Back 33

WARM UP 10

Name: ____________

Amy has a coupon for $140. If she buys a CD player for $65, how much will she have left to spend?

1. a. Draw jumps to show how you could start at 65 and count on to 140. *

+70 +5

40 50 60 70 80 90 100 110 120 130 140

Think of another way you could figure out how much is left.

b. Draw jumps again to show another way. *

-5 -60

40 50 60 70 80 90 100 110 120 130 140

c. Complete this sentence.

140 – 65 = 75 **because** 65 + 75 = 140

2. Suppose the CD player was $78. Write a number sentence to show how much she would have left to spend.

$140 – $78 = $62

34 Count On

© ORIGO Publications Permission is given to instructors to reproduce this page for classroom use.

WORK OUT 10

Name: ____________

1. Figure out each of these in your head. Write the answer. Draw jumps to show how you did it.

a. 160 – 72 = 88 *

70 80 90 100 110 120 130 140 150 160

b. 230 – 68 = 162 *

160 170 180 190 200 210 220 230 240 250

2. Look at these sale items.

$89 $98 $69 $75 $117 $122

For each coupon, write how much would be left to spend after buying each item.

Gift Coupon	skateboard	watch	radio controled car
$120	$ 45	$ 51	$ 31

Gift Coupon	in-line skates	camera	stereo
$150	$ 28	$ 52	$ 33

© ORIGO Publications Permission is given to instructors to reproduce this page for classroom use.

Count On 35

* Answers will vary. This is one example.

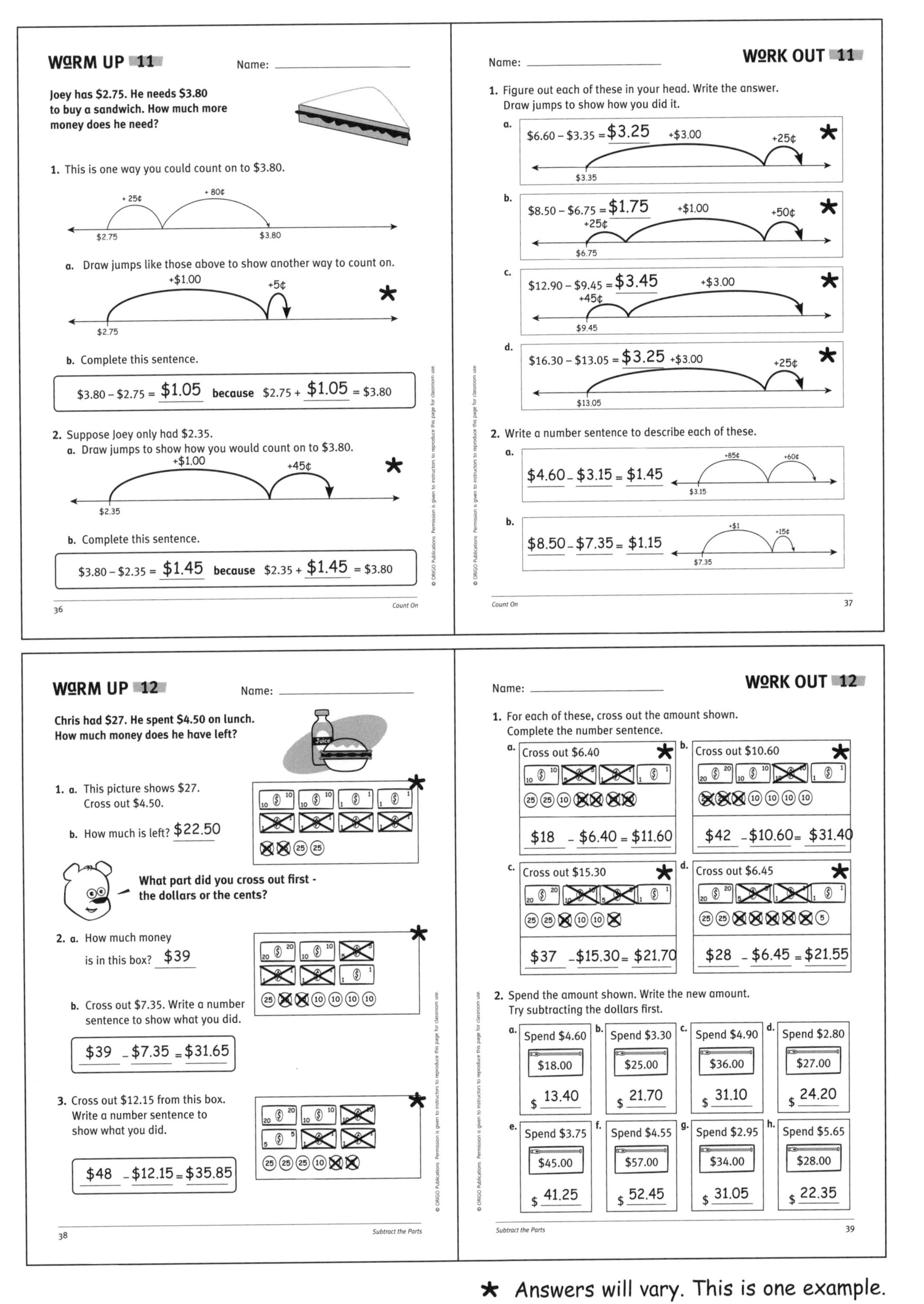

WORM UP 11

Name: ____________

Joey has $2.75. He needs $3.80 to buy a sandwich. How much more money does he need?

1. This is one way you could count on to $3.80.

   + 25¢ + 80¢ — $2.75 — $3.80

   a. Draw jumps like those above to show another way to count on.

   +$1.00 +5¢ * — $2.75

   b. Complete this sentence.

   $3.80 – $2.75 = $1.05 **because** $2.75 + $1.05 = $3.80

2. Suppose Joey only had $2.35.

   a. Draw jumps to show how you would count on to $3.80.

   +$1.00 +45¢ * — $2.35

   b. Complete this sentence.

   $3.80 – $2.35 = $1.45 **because** $2.35 + $1.45 = $3.80

36 — *Count On*

WORK OUT 11

Name: ____________

1. Figure out each of these in your head. Write the answer. Draw jumps to show how you did it.

   a. $6.60 – $3.35 = $3.25 — +$3.00 +25¢ * — $3.35

   b. $8.50 – $6.75 = $1.75 — +25¢ +$1.00 +50¢ * — $6.75

   c. $12.90 – $9.45 = $3.45 — +45¢ +$3.00 * — $9.45

   d. $16.30 – $13.05 = $3.25 — +$3.00 +25¢ * — $13.05

2. Write a number sentence to describe each of these.

   a. $4.60 – $3.15 = $1.45 — +85¢ +60¢ — $3.15

   b. $8.50 – $7.35 = $1.15 — +$1 +15¢ — $7.35

Count On — 37

WORM UP 12

Name: ____________

Chris had $27. He spent $4.50 on lunch. How much money does he have left?

1. a. This picture shows $27. Cross out $4.50. *

   b. How much is left? $22.50

What part did you cross out first - the dollars or the cents?

2. a. How much money is in this box? $39 *

   b. Cross out $7.35. Write a number sentence to show what you did.

   $39 – $7.35 = $31.65

3. Cross out $12.15 from this box. Write a number sentence to show what you did. *

   $48 – $12.15 = $35.85

38 — *Subtract the Parts*

WORK OUT 12

Name: ____________

1. For each of these, cross out the amount shown. Complete the number sentence.

   a. Cross out $6.40 * — $18 – $6.40 = $11.60

   b. Cross out $10.60 * — $42 – $10.60 = $31.40

   c. Cross out $15.30 * — $37 – $15.30 = $21.70

   d. Cross out $6.45 * — $28 – $6.45 = $21.55

2. Spend the amount shown. Write the new amount. Try subtracting the dollars first.

   a. Spend $4.60 — $18.00 — $ 13.40

   b. Spend $3.30 — $25.00 — $ 21.70

   c. Spend $4.90 — $36.00 — $ 31.10

   d. Spend $2.80 — $27.00 — $ 24.20

   e. Spend $3.75 — $45.00 — $ 41.25

   f. Spend $4.55 — $57.00 — $ 52.45

   g. Spend $2.95 — $34.00 — $ 31.05

   h. Spend $5.65 — $28.00 — $ 22.35

Subtract the Parts — 39

* Answers will vary. This is one example.

WORM UP 13

Name: ____________

It is 69.4 miles to the beach by the scenic road and only 32.2 miles by the highway. How much shorter is the highway?

Scenic Road
Highway

1. a. Write the problem. 69.4 - 32.2

What is an easy way to figure out problems such as this?

b. Write the answer. 37.2

How can you check that you are correct?

2. If it was 68.7 miles by the scenic road and 46.3 miles by the highway, how much shorter would the highway be?

Look for an easy way to figure out the answer.

Write the number sentence. 68.7 - 46.3 = 22.4

3. Write two subtraction sentences involving tenths that you could solve the same way. *

a. 76.8 - 34.5 = 42.3 b. 59.6 - 26.2 = 33.4

WORK OUT 13

Name: ____________

1. a. Suppose you had to figure out 47.6 – 23.4 in your head. Write how you would do it. *
First I would take away the whole numbers, 47.6 - 23 equals 24.6. Then I'd subtract the fraction. 24.6 - 0.4 = 24.2.

b. Describe another way you could do it. *
4 tens - 2 tens = 2 tens, then 7 ones - 3 ones is 4 ones, and 6 tenths take 4 tenths is 2 tenths. That's 24.2.

c. Write three subtraction problems you could solve using either of these two methods. *
68.2 - 25.1 | 76.4 - 33.1 | 54.9 - 41.8

2. Use the method you prefer to calculate each of these.

a. 75.9 – 22.5 = 53.4 b. 86.8 – 63.4 = 23.4
c. 34.7 – 13.4 = 21.3 d. 59.3 – 47.1 = 12.2
e. 256.4 – 132.3 = 124.1 f. 563.7 – 241.6 = 322.1

3. Write the missing number in each pyramid. The sum of two numbers must equal the number above. (Example: 67.5 above 34.1, 33.4)

a. 87.9; 43.5, 44.4; 12.2, 31.3, 13.1
b. 96.8; 43.5, 53.3; 12.3, 31.2, 22.1
c. 78.9; 35.6, 43.3; 22.4, 13.2, 30.1

WORM UP 14

Name: ____________

Ryan had $158 in savings. He bought new basketball shoes for $49. How much money does he have left?

How could you use 158 – 50 to help you figure out 158 – 49?

1. Describe your method. Complete this sentence to help you.
158 – 50 = 108 so 158 – 49 = 109

2. What is an easy way to figure out 128 – 59 in your head? Complete this sentence to help you.
128 – 60 = 68 so 128 – 59 = 69

Describe another strategy you could use.

3. a. Suppose you had to figure out 228 – 39. Write an easier number sentence you know would help.
228 - 40 = 188

b. Write the answer.
228 – 39 = 189

WORK OUT 14

Name: ____________

1. For each of these, write an easier number sentence that will help you figure out the answer. Write the answer. *

a. 258 - 40 = 218 so 258 – 39 = 219
b. 177 - 40 = 137 so 177 – 38 = 139
c. 265 - 50 = 215 so 265 – 47 = 218
d. 138 - 60 = 78 so 138 – 59 = 79
e. 187 - 70 = 117 so 187 – 69 = 118
f. 268 - 100 = 168 so 268 – 99 = 169

2. Write the answers. Place a ✓ above the numbers you adjusted. *

a. 167 - 49 = 118 b. 348 – 69 = 279 c. 236 – 78 = 158

3. Take the numbers on the spokes away from the number in the center. Write the answers around the outside.

a. Center 296: –19 → 277; –38 → 258; –69 → 227; –77 → 219; –58 → 238; –18 → 278; –49 → 247; –28 → 268

b. Center 187: –38 → 149; –48 → 139; –29 → 158; –69 → 118; –58 → 129; –39 → 148; –49 → 138; –19 → 168

* Answers will vary. This is one example.

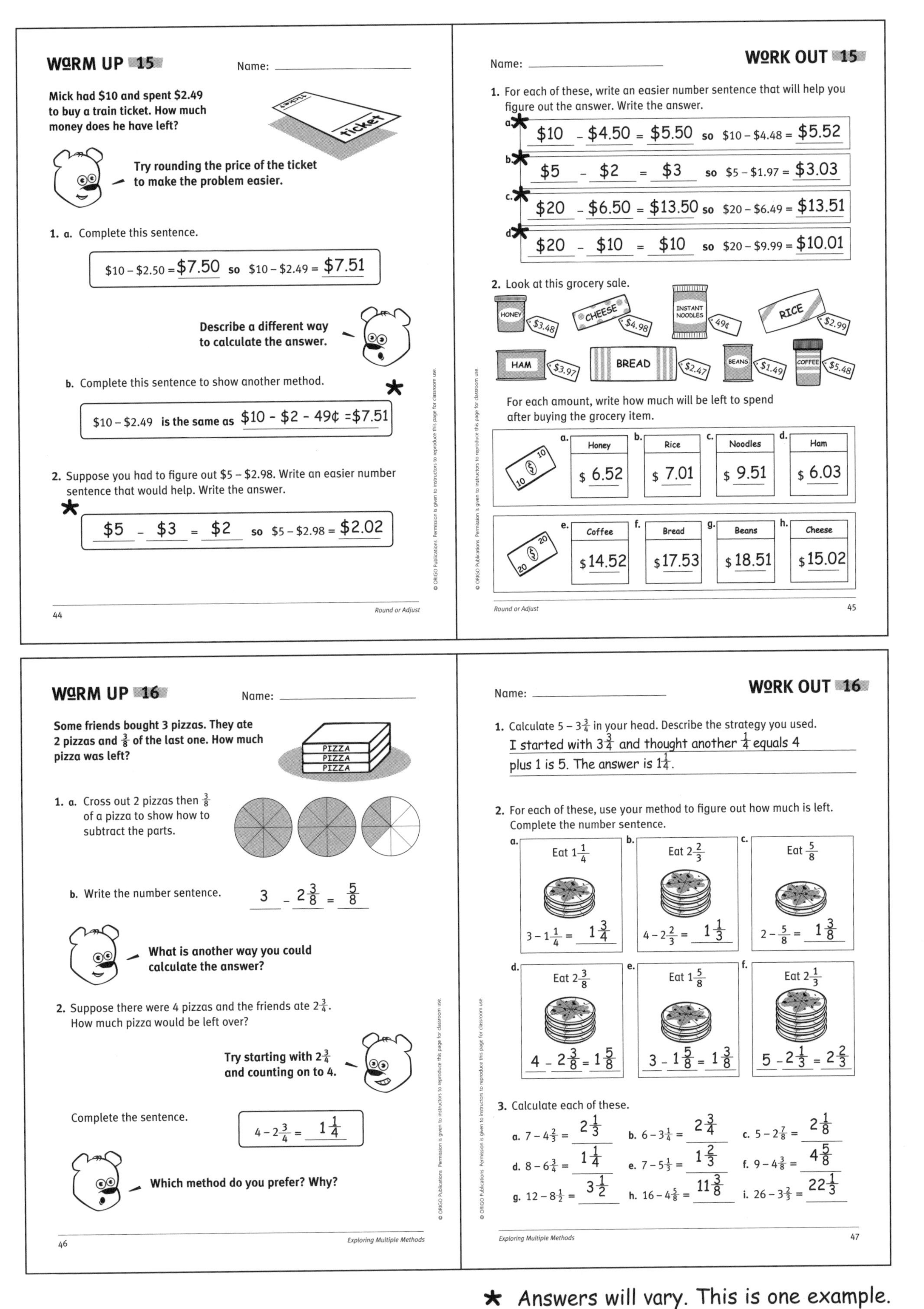

WARM UP 15

Name: ________________

Mick had $10 and spent $2.49 to buy a train ticket. How much money does he have left?

Try rounding the price of the ticket to make the problem easier.

1. a. Complete this sentence.

$10 – $2.50 = $7.50 so $10 – $2.49 = $7.51

Describe a different way to calculate the answer.

b. Complete this sentence to show another method. ★

$10 – $2.49 is the same as $10 - $2 - 49¢ = $7.51

2. Suppose you had to figure out $5 – $2.98. Write an easier number sentence that would help. Write the answer.

★ $5 – $3 = $2 so $5 – $2.98 = $2.02

44 — *Round or Adjust*

WORK OUT 15

Name: ________________

1. For each of these, write an easier number sentence that will help you figure out the answer. Write the answer.

a. ★ $10 – $4.50 = $5.50 so $10 – $4.48 = $5.52

b. ★ $5 – $2 = $3 so $5 – $1.97 = $3.03

c. ★ $20 – $6.50 = $13.50 so $20 – $6.49 = $13.51

d. ★ $20 – $10 = $10 so $20 – $9.99 = $10.01

2. Look at this grocery sale.

Honey $3.48, Cheese $4.98, Instant Noodles 49¢, Rice $2.99, Ham $3.97, Bread $2.47, Beans $1.49, Coffee $5.48

For each amount, write how much will be left to spend after buying the grocery item.

$10:

a. Honey	b. Rice	c. Noodles	d. Ham
$ 6.52	$ 7.01	$ 9.51	$ 6.03

$20:

e. Coffee	f. Bread	g. Beans	h. Cheese
$14.52	$17.53	$18.51	$15.02

Round or Adjust — 45

© ORIGO Publications Permission is given to instructors to reproduce this page for classroom use.

WARM UP 16

Name: ________________

Some friends bought 3 pizzas. They ate 2 pizzas and $\frac{3}{8}$ of the last one. How much pizza was left?

PIZZA PIZZA PIZZA

1. a. Cross out 2 pizzas then $\frac{3}{8}$ of a pizza to show how to subtract the parts.

b. Write the number sentence. $3 - 2\frac{3}{8} = \frac{5}{8}$

What is another way you could calculate the answer?

2. Suppose there were 4 pizzas and the friends ate $2\frac{3}{4}$. How much pizza would be left over?

Try starting with $2\frac{3}{4}$ and counting on to 4.

Complete the sentence. $4 - 2\frac{3}{4} = 1\frac{1}{4}$

Which method do you prefer? Why?

46 — *Exploring Multiple Methods*

WORK OUT 16

Name: ________________

1. Calculate $5 - 3\frac{3}{4}$ in your head. Describe the strategy you used.

I started with $3\frac{3}{4}$ and thought another $\frac{1}{4}$ equals 4 plus 1 is 5. The answer is $1\frac{1}{4}$.

2. For each of these, use your method to figure out how much is left. Complete the number sentence.

a. Eat $1\frac{1}{4}$: $3 - 1\frac{1}{4} = 1\frac{3}{4}$

b. Eat $2\frac{2}{3}$: $4 - 2\frac{2}{3} = 1\frac{1}{3}$

c. Eat $\frac{5}{8}$: $2 - \frac{5}{8} = 1\frac{3}{8}$

d. Eat $2\frac{3}{8}$: $4 - 2\frac{3}{8} = 1\frac{5}{8}$

e. Eat $1\frac{5}{8}$: $3 - 1\frac{5}{8} = 1\frac{3}{8}$

f. Eat $2\frac{1}{3}$: $5 - 2\frac{1}{3} = 2\frac{2}{3}$

3. Calculate each of these.

a. $7 - 4\frac{2}{3} = 2\frac{1}{3}$ b. $6 - 3\frac{1}{4} = 2\frac{3}{4}$ c. $5 - 2\frac{7}{8} = 2\frac{1}{8}$

d. $8 - 6\frac{3}{4} = 1\frac{1}{4}$ e. $7 - 5\frac{1}{3} = 1\frac{2}{3}$ f. $9 - 4\frac{3}{8} = 4\frac{5}{8}$

g. $12 - 8\frac{1}{2} = 3\frac{1}{2}$ h. $16 - 4\frac{5}{8} = 11\frac{3}{8}$ i. $26 - 3\frac{2}{3} = 22\frac{1}{3}$

Exploring Multiple Methods — 47

★ Answers will vary. This is one example.

CHECK UP 2

Name: ______________

1. Figure out these in your head. Write the answers.

a. 46.7 – 23.4 = 23.3   b. 243 – 26 = 217   c. 220 – 68 = 152

d. $4 - 2\frac{1}{4} = 1\frac{3}{4}$ e. 177 – 38 = 139   f. $7 - 5\frac{1}{3} = 1\frac{2}{3}$

g. $18 – $4.80 = $13.20   h. $5.70 – $2.35 = $3.35

2. a. Write the answer.

168 – 39 = 129

Think about how you figured it out.

b. Write three number sentences you could solve the same way. *

152 – 38 = 114   142 – 29 = 113   175 – 59 = 116

3. Write the missing numbers in each pyramid.
The sum of two numbers must equal the number above.

(Example: 140 / 59, 81)

a.
$7
$3.48 | $3.52
$1.28 | $2.20 | $1.32

b.
196
79 | 117
45 | 34 | 83

c.
88.9
65.6 | 23.3
53.4 | 12.2 | 11.1

d.
127
68 | 59
39 | 29 | 30

JUST FOR FUN 2

Name: ______________

There were two tourists standing outside the White House. One of them was the father of the other one's son. How could this be?

Solve this riddle by figuring out the answers below. Write the letters above their matching answers at the bottom of the page. Some letters appear more than once.

1. 253 – 27 = 226 **W**
2. 120 – 59 = 61 **B**
3. $4.90 – $2.45 = $2.45 **D**
4. $15 – $3.30 = $11.70 **H**
5. 86.7 – 51.3 = 35.4 **S**
6. 145 – 27 = 118 **R**
7. 150 – 85 = 65 **T**
8. 167 – 39 = 128 **A**
9. $21 – $4.60 = $16.40 **Y**
10. $6.50 – $2.15 = $4.35 **F**
11. $3 - 1\frac{3}{5} = 1\frac{2}{5}$ **N**
12. 75.4 – 23.1 = 52.3 **E**
13. 127 – 59 = 68 **I**
14. $20 – $8.99 = $11.01 **U**

T	H	E	Y
65	$11.70	52.3	$16.40

W	E	R	E
226	52.3	118	52.3

H	U	S	B	A	N	D
$11.70	$11.01	35.4	61	128	$1\frac{2}{5}$	$2.45

A	N	D
128	$1\frac{2}{5}$	$2.45

W	I	F	E
226	68	$4.35	52.3

* Answers will vary. This is one example.

MULTIPLICATION STRATEGIES

Use doubles

3.2 x 4 is the same as 3.2 x 2 x 2

Double and halve

3 x 14 is the same as 6 x 7

$3.50 x 12 is the same as $7 x 6

36 x 25 is the same as 18 x 50 or 9 x 100

Use place value

32 x 11 is the same as (32 x 10) + (32 x 1)

Use compatible pairs

4 x 7 x 5 x 3 is the same as (4 x 5) x (7 x 3)

Use factors

22 x 15 is the same as 11 x 2 x 5 x 3 or (5 x 2) x (11 x 3)

Use division

$\frac{1}{4}$ *x 36 is the same as 36 ÷ 4*

$\frac{1}{3}$ *of 12 is the same as 12 ÷ 3*

$\frac{1}{5}$ *of 250 is the same as* ($\frac{1}{10}$ *of 250) x 2*

Name: ______________________

1. a. Double 3.2 then double again. Write the answers in the circles.

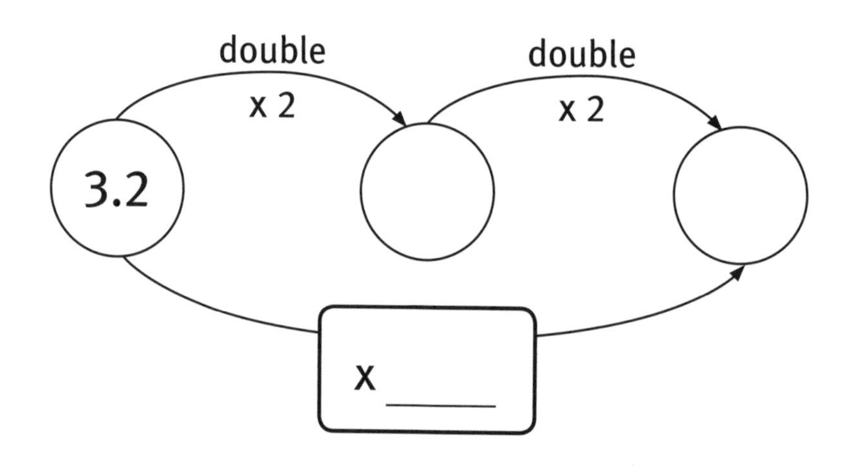

b. Write a number in the box above.
Complete the number sentence
to show the answer.

3.2 x ______ = ______

2. a. Complete this picture.

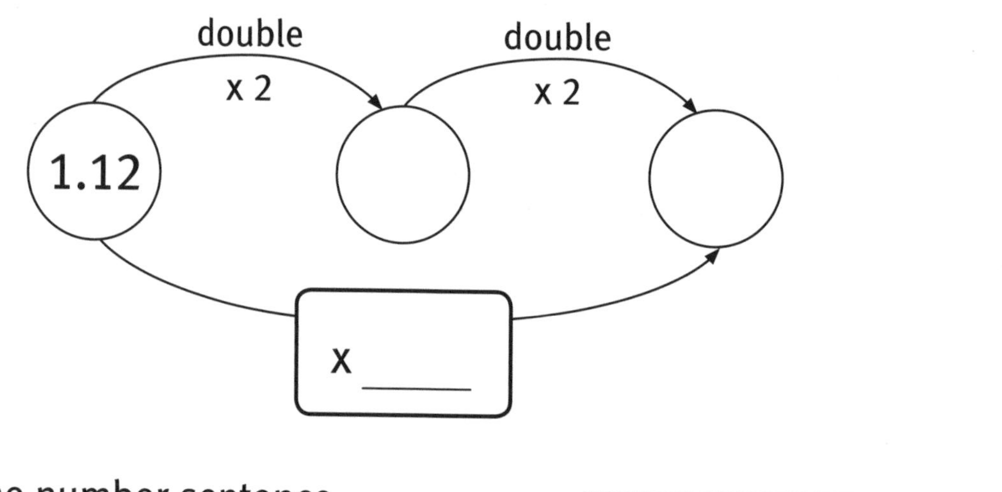

b. Write the number sentence.

1.12 x ______ = ______

3. Write some other number sentences you could solve by thinking double double.

a. ______ x 4 = ______ b. ______ x 4 = ______ c. ______ x 4 = ______

Name: ______________________

1. Complete each picture. Write the number sentence.

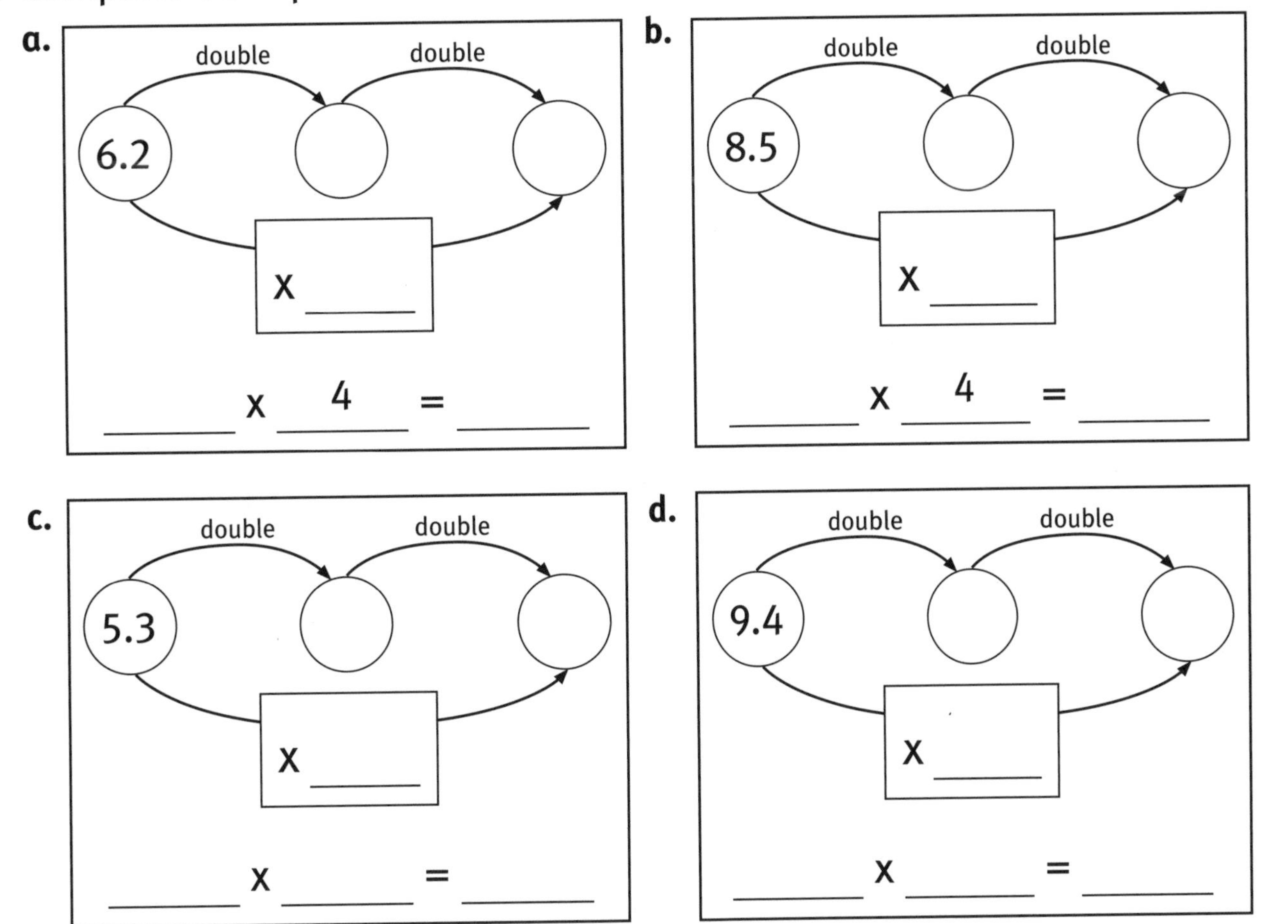

2. Think double double to complete these number sentences.

a. 7.4 x 4 = ______ = 4 x ____ b. 3.15 x 4 = ______ = ____ x ____

3. Complete each picture. Write the number sentence.

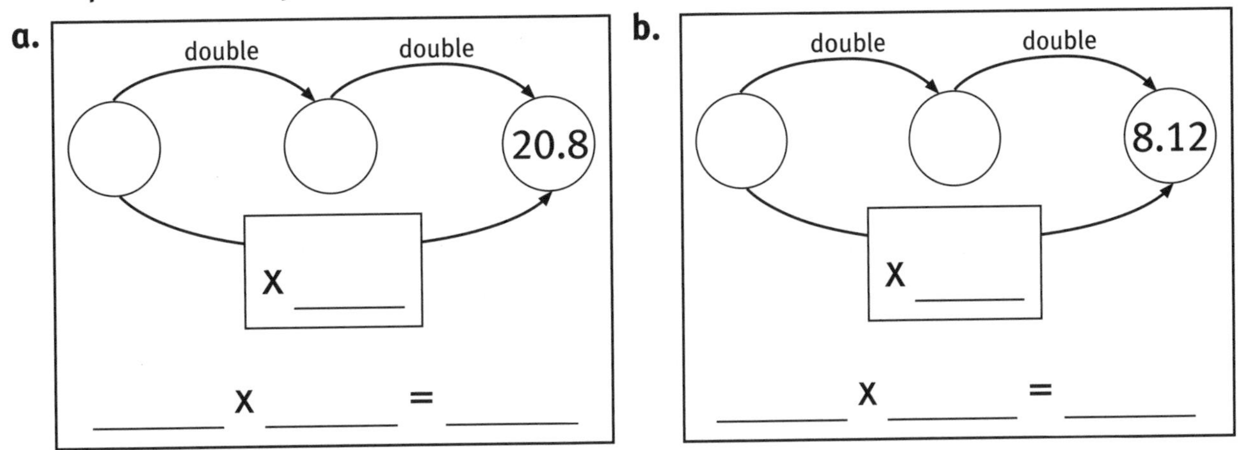

WoRM UP 18

Name: ____________________

1. a. Write numbers to show the dimensions of this grid.

____ x ____

b. Write numbers to show how you would calculate the total number of squares.

____ x ____

2. a. Suppose we cut the grid in half and rearranged the pieces as shown.

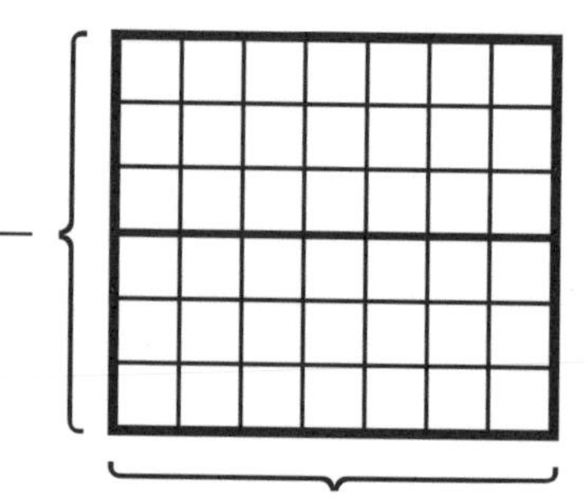

Has the total number of squares changed? ________

b. Write the new dimensions.

c. Complete the number sentence used to calculate the total number of squares.

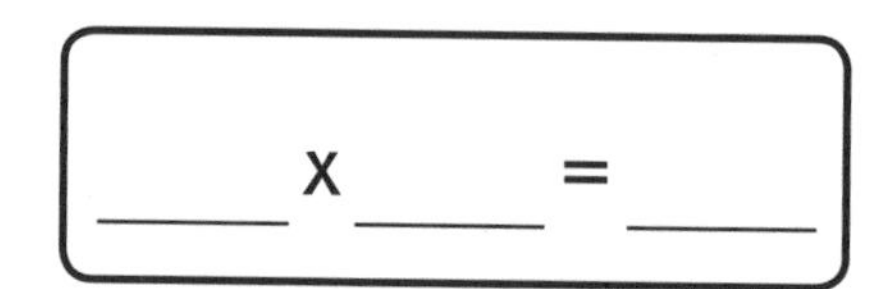

d. Complete this sentence.

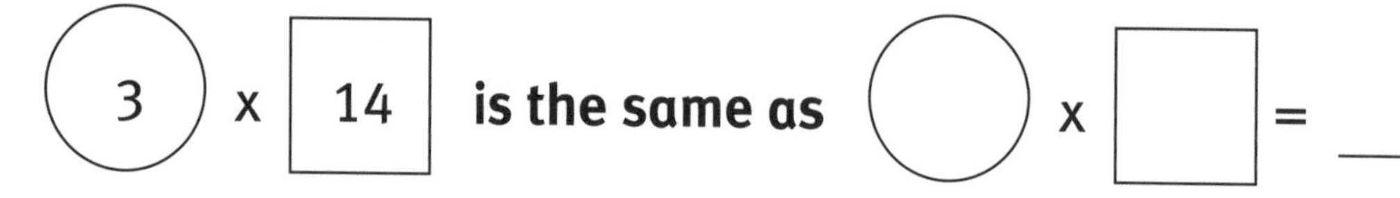

Compare the numbers in the circles, then the numbers in the boxes. What do you notice?

Name: ____________________

1. For each of these, double one number and halve the other to make a new number sentence that is easier to calculate. Write the answer.

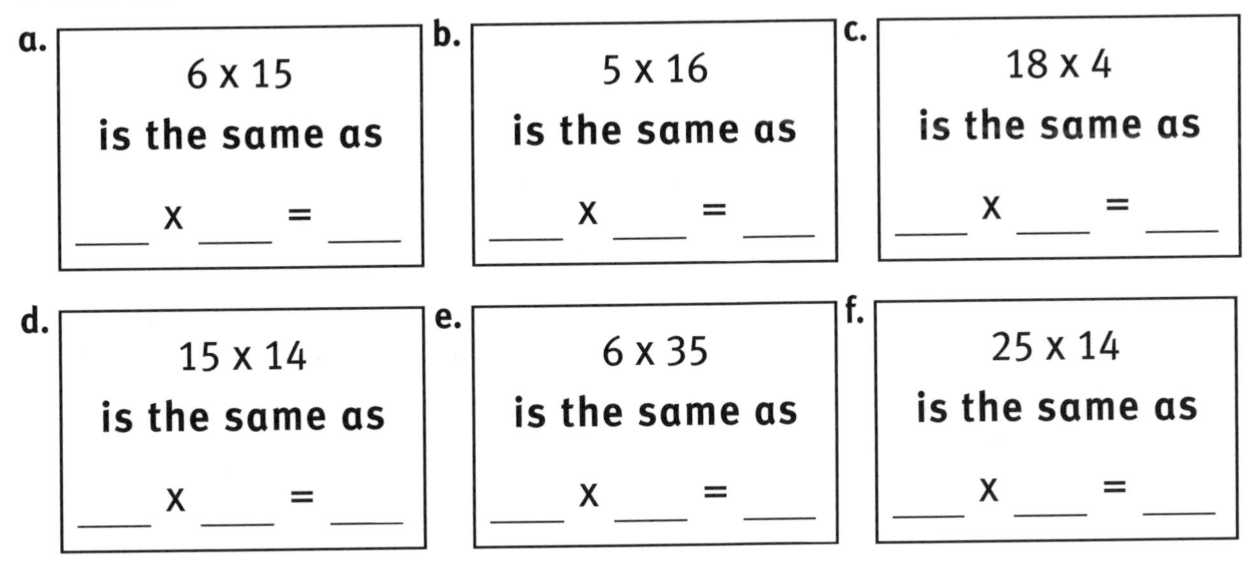

2. For each of these, draw an arrow to the fact you could use to help figure it out. Write the answer.

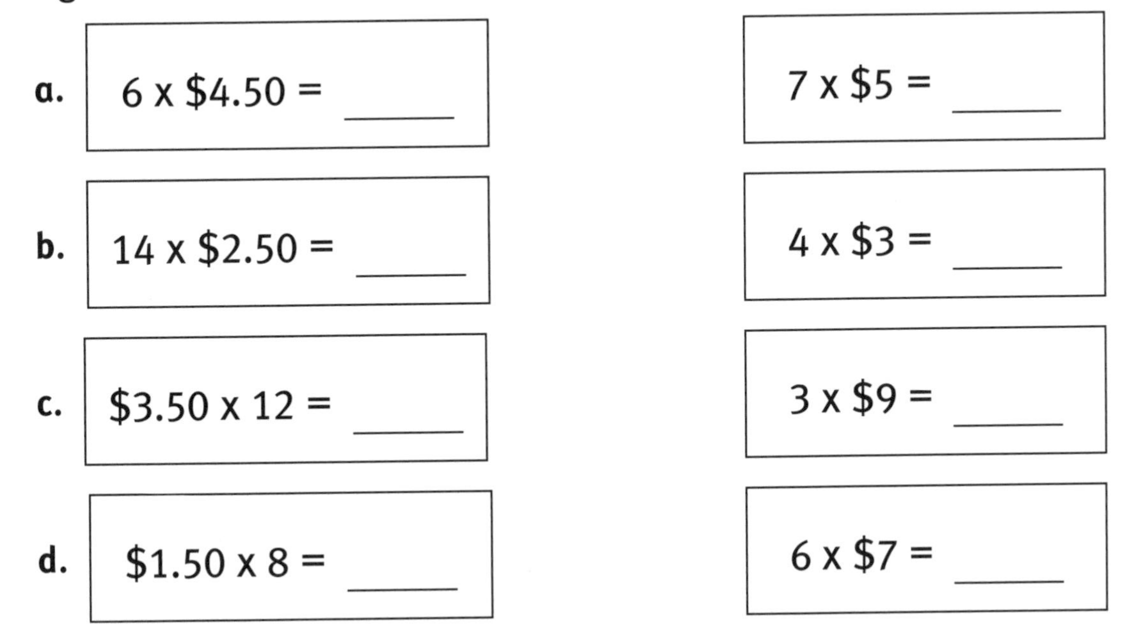

3. Write some other number sentences you could solve using this strategy.

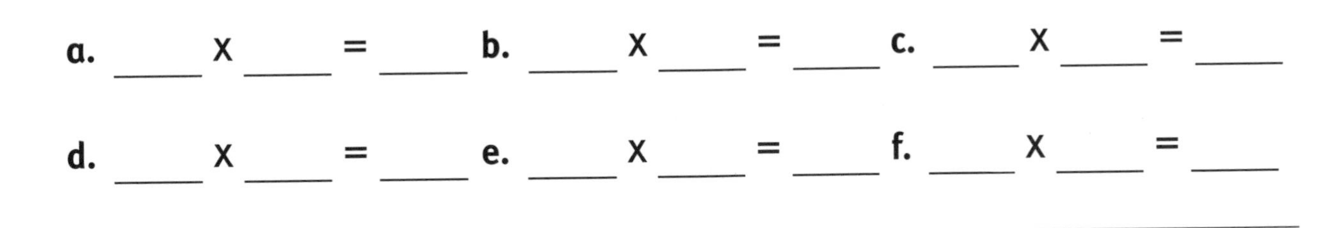

Name: ____________________

A farmer planted 36 rows of corn.
Each row had 25 plants.
How many plants are there in all?

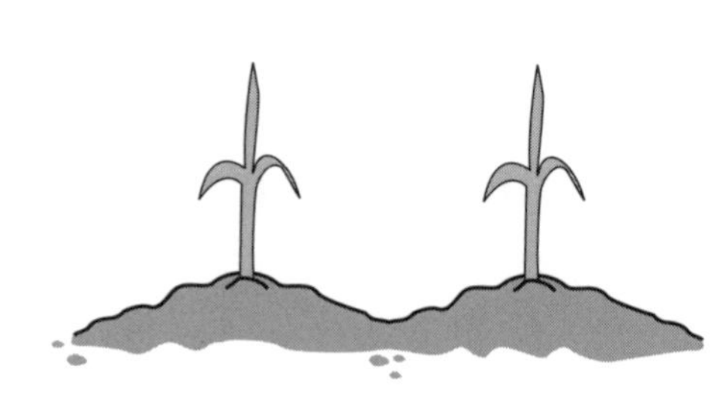

1. a. Double the first number and halve the other number to make an equivalent problem.

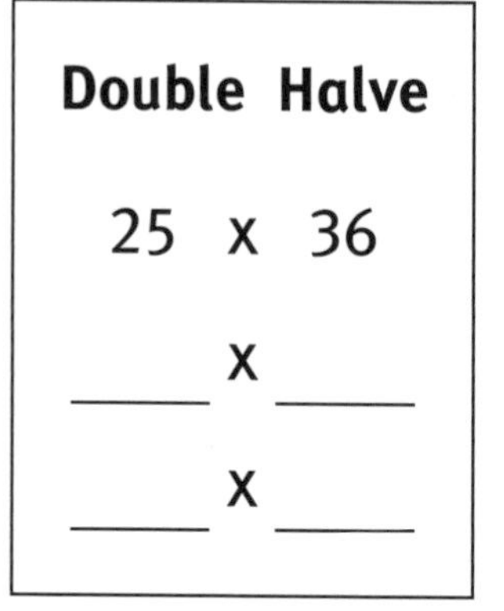

b. Repeat the step until you make a problem that is easier to solve.

c. Complete this sentence.

25 x 36 **is the same as** ____ x ____ = ____

2. a. Use the doubling and halving strategy again to solve 24 x 15.

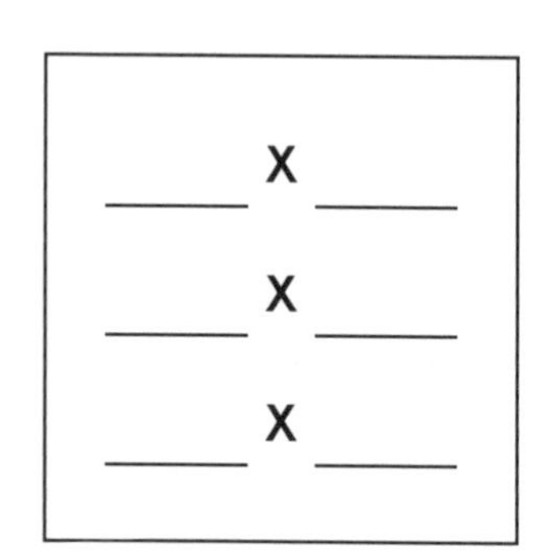

b. Complete this sentence.

24 x 15 **is the same as** ____ x ____ = ____

Name: ______________________

WORK OUT 19

1. For each of these, double one number and halve the other to make a problem that is easier to solve. Write the answer.

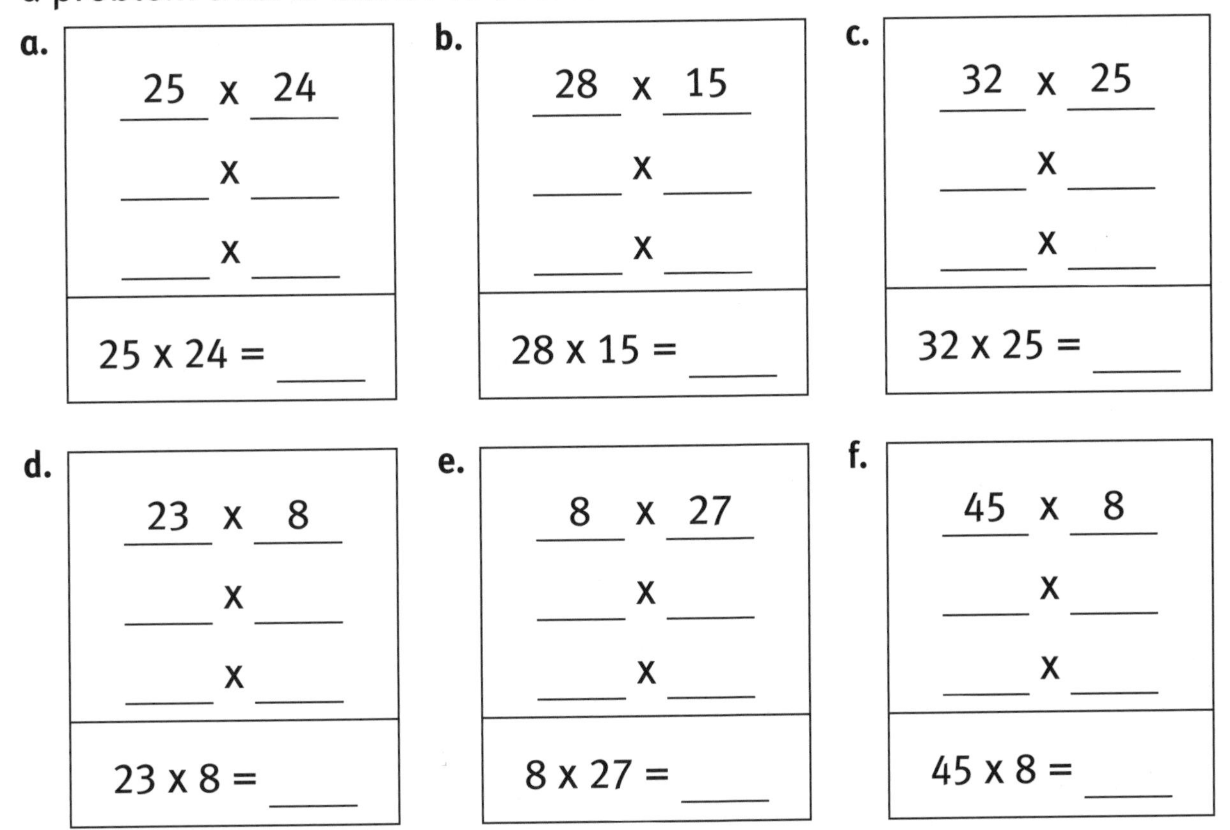

2. For each of these, draw an arrow to the problem you could use to help figure it out. Write the answer.

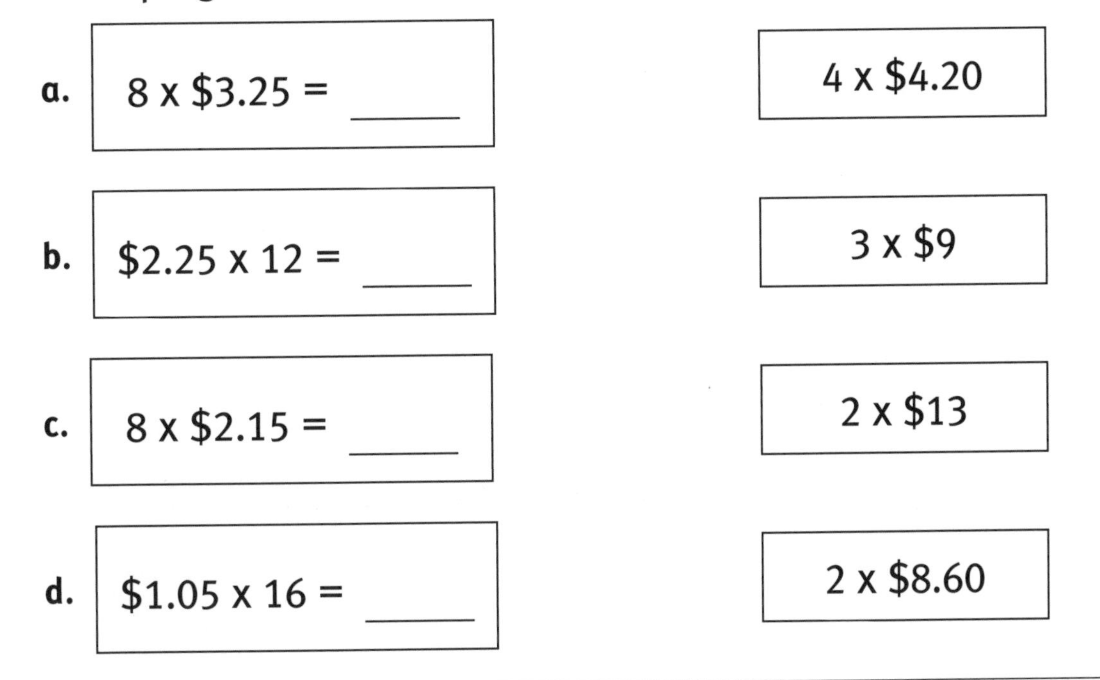

Name: ______________________

A school reserved 11 rows of 32 seats at a baseball game. How many seats were reserved?

Multiply 32 by the tens then the ones.

1. a. Complete this sentence to show one method.

32 x 11 **is the same as** (32 x _____) + (32 x _____)

b. Write the answer. _______

2. a. Suppose there were 28 seats in each row.
How many seats were reserved? Complete this sentence.

28 x 11 **is the same as** (_____ x _____) + (_____ x _____)

b. Write the answer. _______

3. Figure out 13 x 21 in your head. Complete this sentence.

13 x 21 **is the same as** (_____ x _____) + (_____ x _____)

What is the answer?

Name: ______________________

1. Multiply the tens then the ones to figure out each cost.
Complete the sentence.

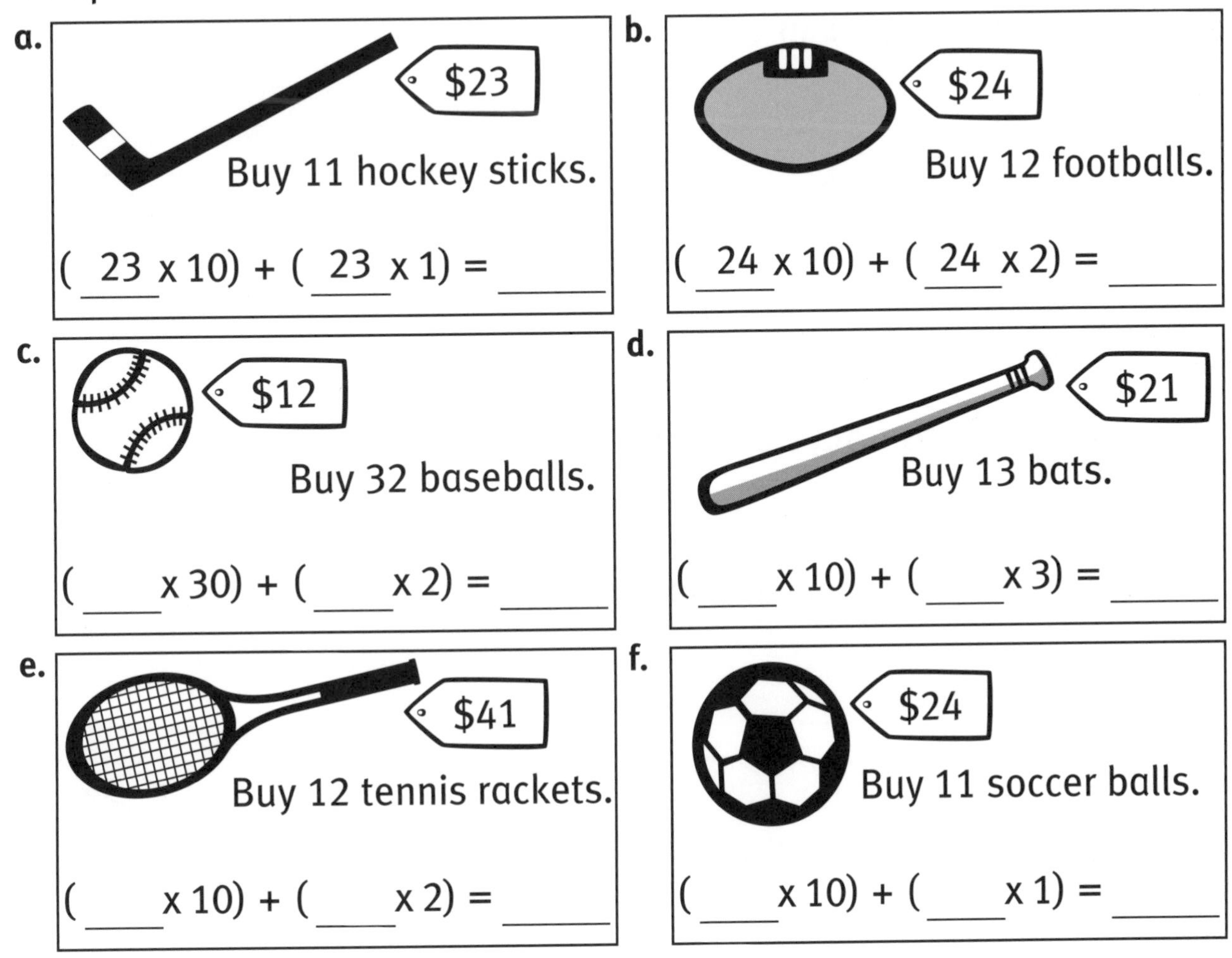

2. Complete each sentence.

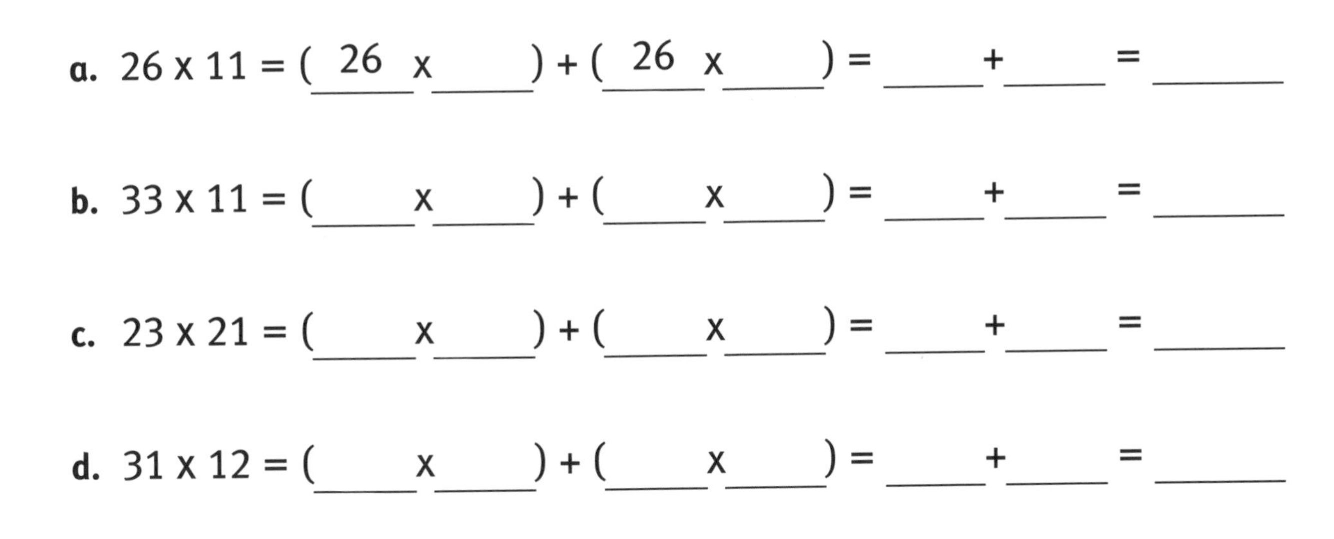

a. 26 x 11 = (26 x ____) + (26 x ____) = ____ + ____ = ____

b. 33 x 11 = (____ x ____) + (____ x ____) = ____ + ____ = ____

c. 23 x 21 = (____ x ____) + (____ x ____) = ____ + ____ = ____

d. 31 x 12 = (____ x ____) + (____ x ____) = ____ + ____ = ____

Name: ____________________

There are 4 layers of boxes. Each layer has 7 rows of 5 boxes. Each box has 3 cans. How many cans in all?

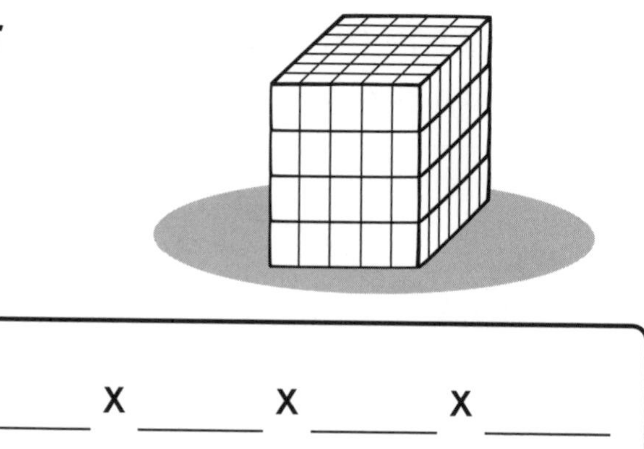

1. **a.** Write the numbers.

_____ x _____ x _____ x _____

Look for an easy way to figure out the answer.

b. Draw ⤻ to connect the numbers that work well together.

c. Write the total. _____

How did you figure it out in your head?

2. Suppose there were 3 layers of boxes. If each layer had 5 rows of 7 boxes and each box had 8 cans, how many cans in all?

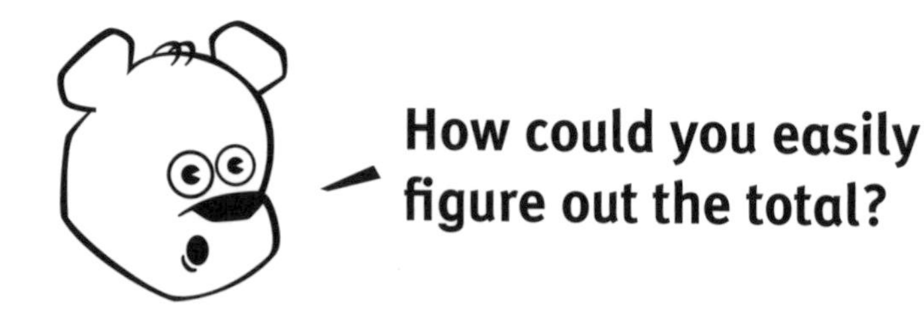

Write the numbers in the order that you would multiply then write the answer.

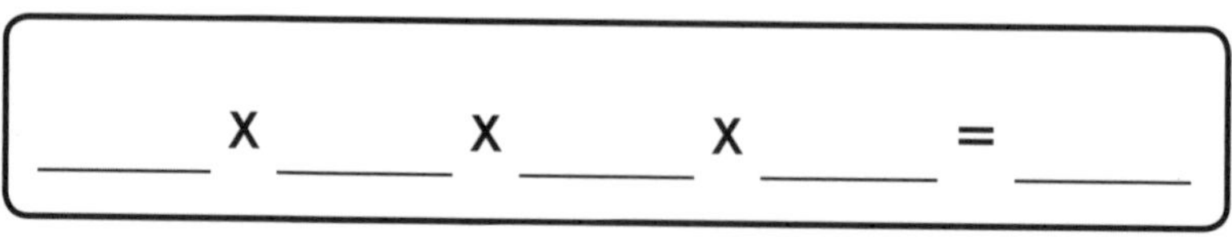

Name: ______________________

1. For each of these, draw ↶↷ to connect the pairs you would multiply together first. Write the total.

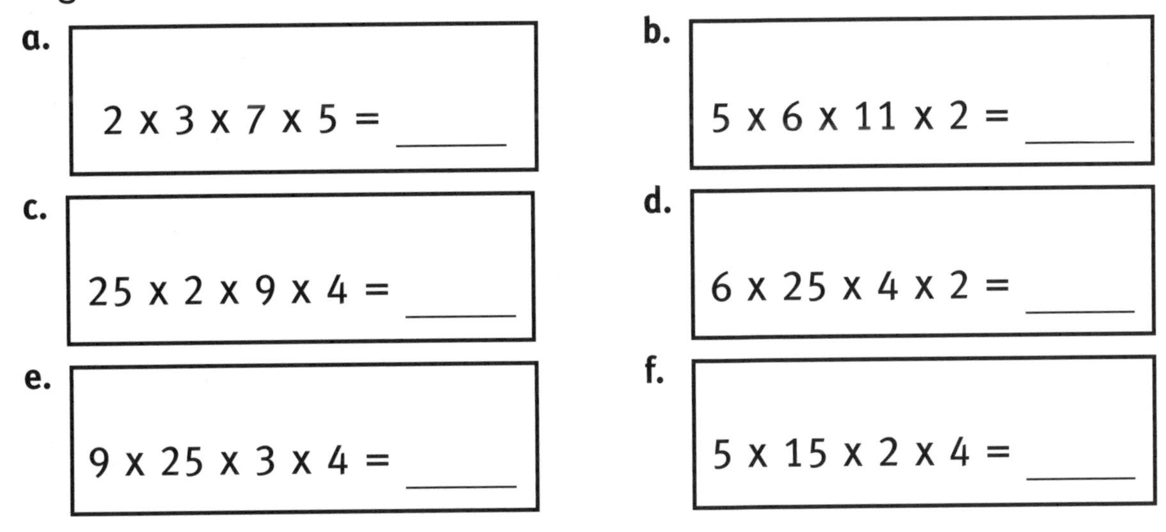

2. Use the same strategy to calculate the answers to these.

a. 5 x 5 x 15 x 4 = ____

b. 5 x 11 x 8 x 5 = ____

c. 15 x 4 x 2 x 25 = ____

d. 2 x 11 x 2 x 25 = ____

e. 2 x 3 x 75 x 2 = ____

f. 6 x 5 x 7 x 0 = ____

3. Write pairs of numbers that work well together.

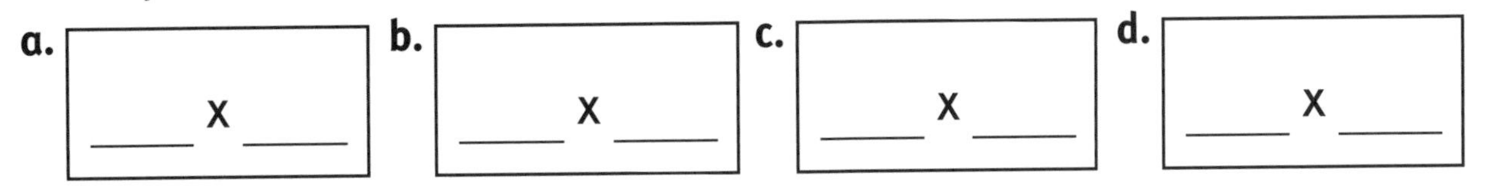

4. Write a number greater than one in each space. Make it easy to figure out the answer in your head. Write the answer.

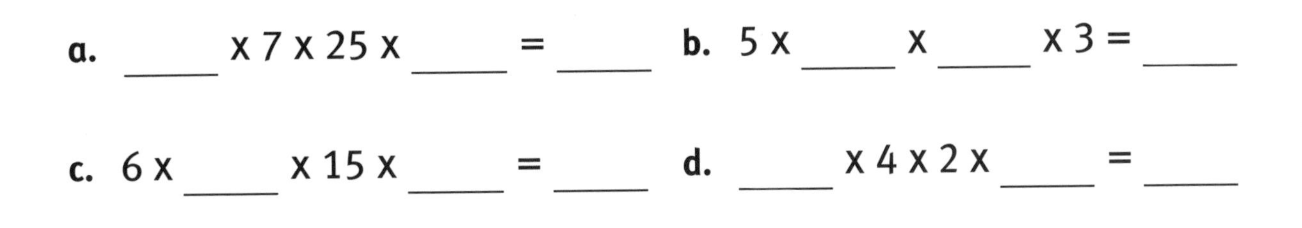

WARM UP 22

Name: ____________________

A class of 22 students is going to a football match. Tickets cost $15 each. What is the total cost of the tickets?

Try using the factors of 22 and 15.

1. a. Complete each of these.

22 **is the same as** ____ x ____	15 **is the same as** ____ x ____

b. Write a new sentence then look for an easy way to figure out the answer. Write the answer.

22 x 15 **is the same as** ____ x ____ x ____ x ____ = ____

2. Suppose there were only 18 students in the class.

a. Complete these sentences to show factors of 18.

18 **is the same as** ____ x ____

18 **is the same as** ____ x ____

b. Choose a pair of factors that will make it easy to figure out the total. Complete this sentence.

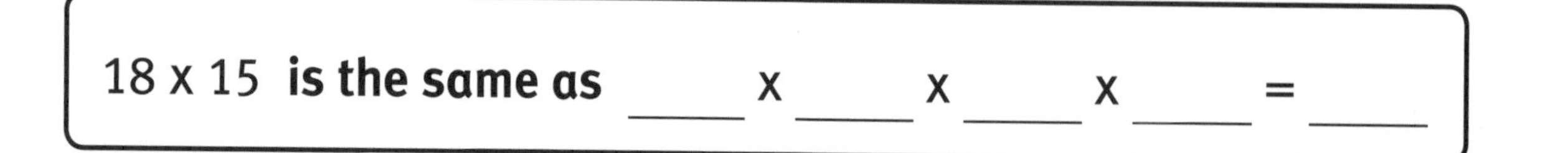

18 x 15 **is the same as** ____ x ____ x ____ x ____ = ____

Name: ______________________

1. Complete each of these to show factors greater than one.

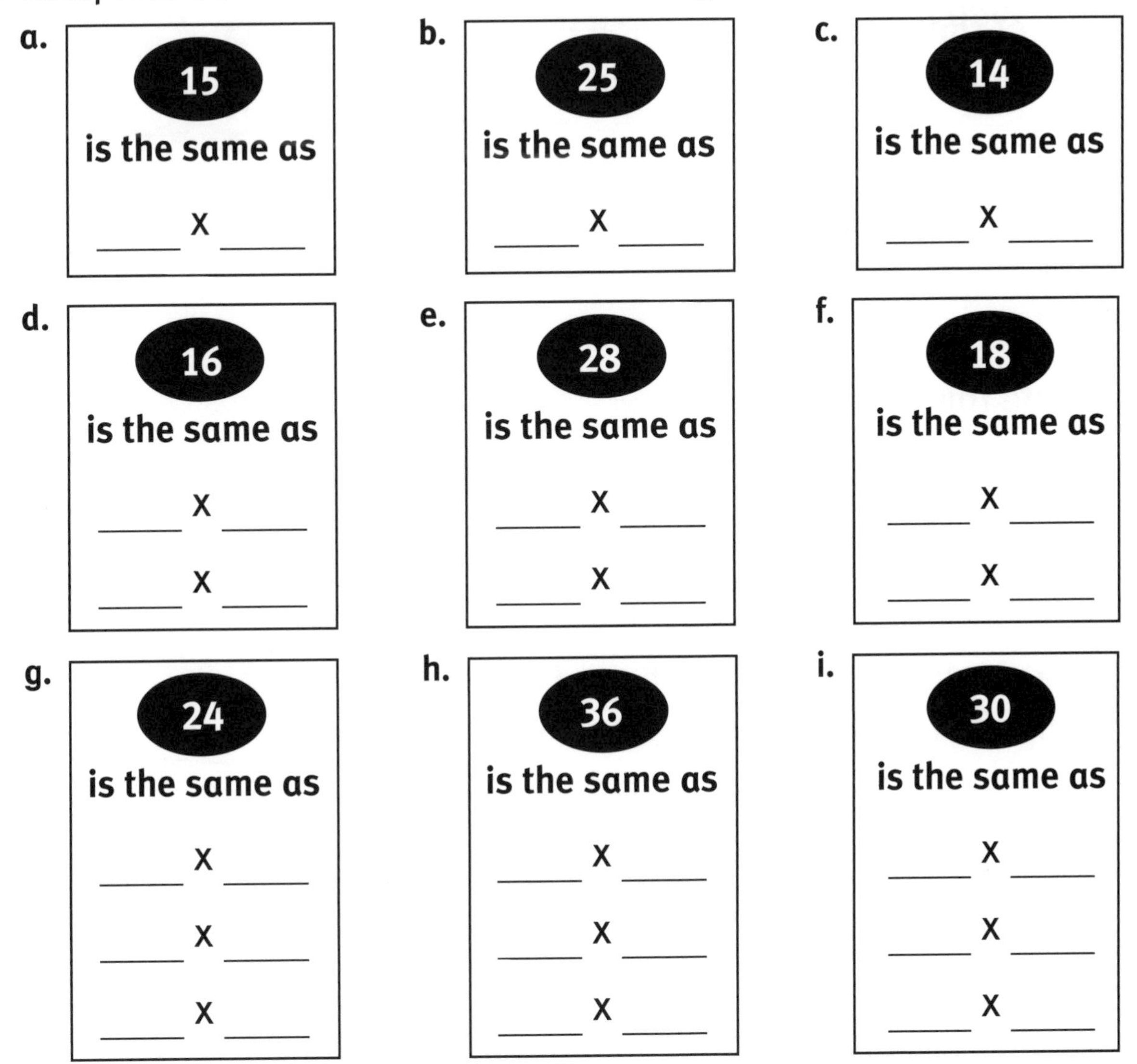

2. For each of these, break each number into two factors to make it easier to figure out. Complete the number sentence.

a.

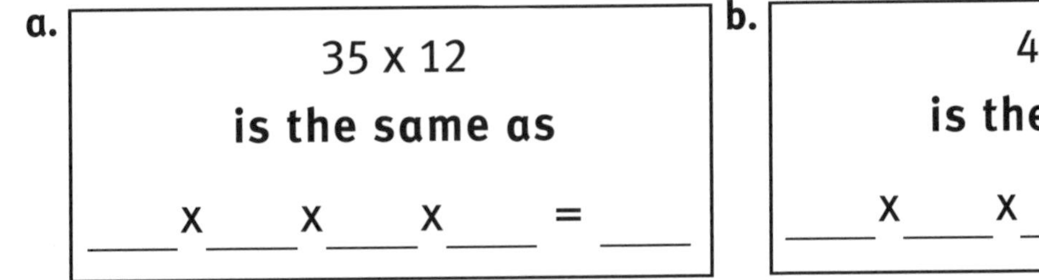

b.

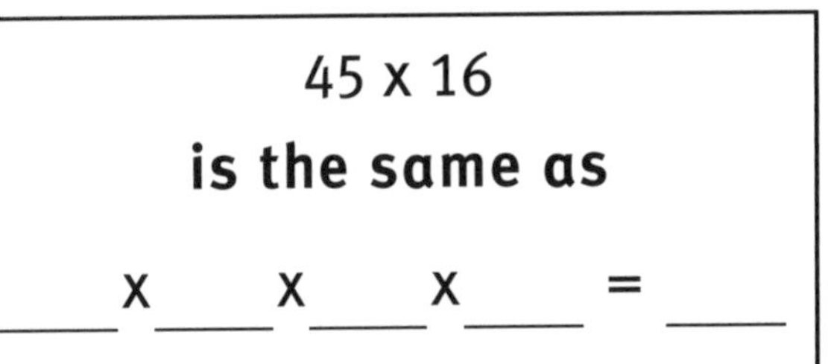

3. Use the same strategy to figure out each of these.

a. 15 x 24 = ______ b. 16 x 25 = ______ c. 28 x 15 = ______

Name: ____________________

Judy bought a bag of 24 candy pieces.
She ate $\frac{1}{2}$ of the bag of candy.
How many candy pieces did she eat?

How can you use division to solve $\frac{1}{2}$ of 24?

1. Complete this sentence.

$\frac{1}{2}$ of 24 **is the same as** 24 ÷ ______ = ______

2. a. Suppose Judy ate $\frac{1}{4}$ of the bag of candy.
Complete this sentence to show how many she ate.

$\frac{1}{4}$ of 24 **is the same as** 24 ÷ ______ = ______

How did you figure out 24 ÷ 4?

b. Complete this sentence to show one way to do it.

______ x ______ = 24 **so** 24 ÷ ______ = ______

3. Use the same strategy to figure out these in your head.

a. $\frac{1}{2}$ of 48 = ______

b. $\frac{1}{4}$ of 36 = ______

Name: ______________________

1. Use division to help you figure out each of these.
Write the missing numbers.

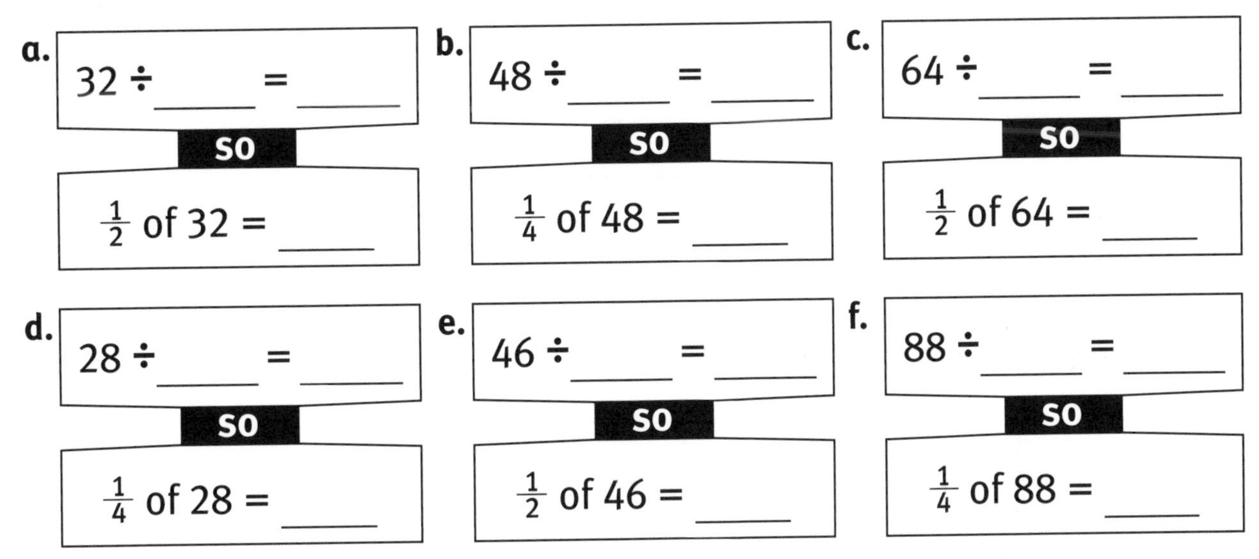

a. 32 ÷ ____ = ____ **so** $\frac{1}{2}$ of 32 = ____

b. 48 ÷ ____ = ____ **so** $\frac{1}{4}$ of 48 = ____

c. 64 ÷ ____ = ____ **so** $\frac{1}{2}$ of 64 = ____

d. 28 ÷ ____ = ____ **so** $\frac{1}{4}$ of 28 = ____

e. 46 ÷ ____ = ____ **so** $\frac{1}{2}$ of 46 = ____

f. 88 ÷ ____ = ____ **so** $\frac{1}{4}$ of 88 = ____

2. Figure out each of these. Write the division sentence.

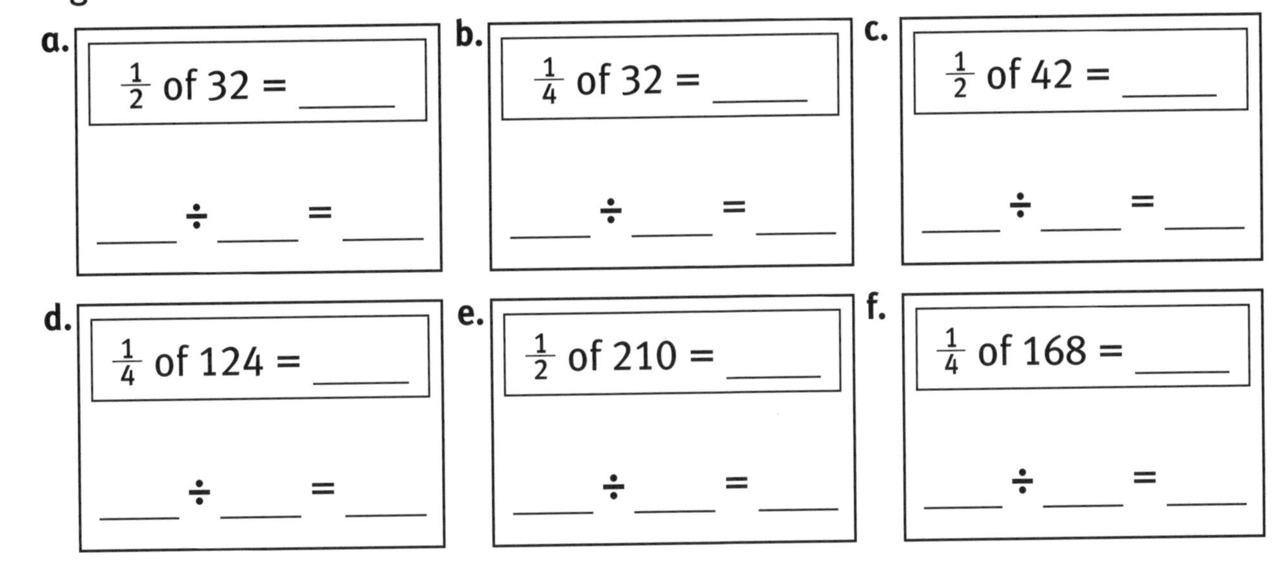

a. $\frac{1}{2}$ of 32 = ____
____ ÷ ____ = ____

b. $\frac{1}{4}$ of 32 = ____
____ ÷ ____ = ____

c. $\frac{1}{2}$ of 42 = ____
____ ÷ ____ = ____

d. $\frac{1}{4}$ of 124 = ____
____ ÷ ____ = ____

e. $\frac{1}{2}$ of 210 = ____
____ ÷ ____ = ____

f. $\frac{1}{4}$ of 168 = ____
____ ÷ ____ = ____

3. Calculate the answers to these. Use division to help.

	248	440	600	880	148	136
$\frac{1}{2}$ of						
$\frac{1}{4}$ of						

WARM UP 24

Name: ______________________

Matt dropped a carton of 12 eggs.
One third of the eggs were broken.
How many were broken?

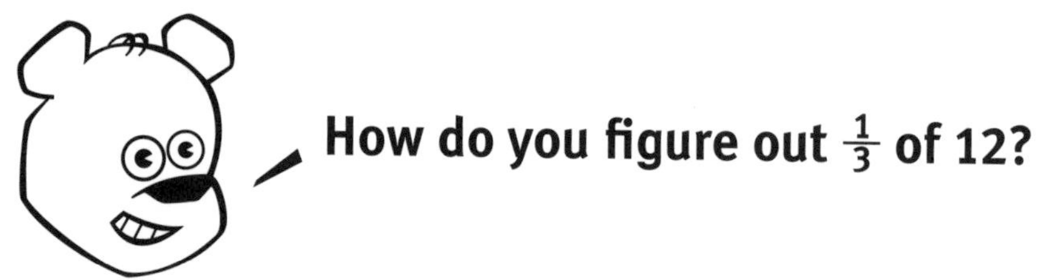

1. a. Complete this sentence.

$\frac{1}{3}$ of 12 **is the same as** 12 ÷ _____ = _____

How could you use multiplication to figure out $\frac{1}{3}$ of 12?

b. Complete this sentence.

_____ x _____ = 12 **so** $\frac{1}{3}$ of 12 = _____

2. Suppose $\frac{1}{6}$ of the eggs were broken.
Complete this sentence to show how many were broken.

$\frac{1}{6}$ of 12 **is the same as** 12 ÷ _____ = _____

3. Use the same strategy to figure out these in your head.

a. $\frac{1}{3}$ of 27 = _____

b. $\frac{1}{6}$ of 48 = _____

Name: ______________________

1. Use division to help you figure out each of these.
   Write the missing numbers.

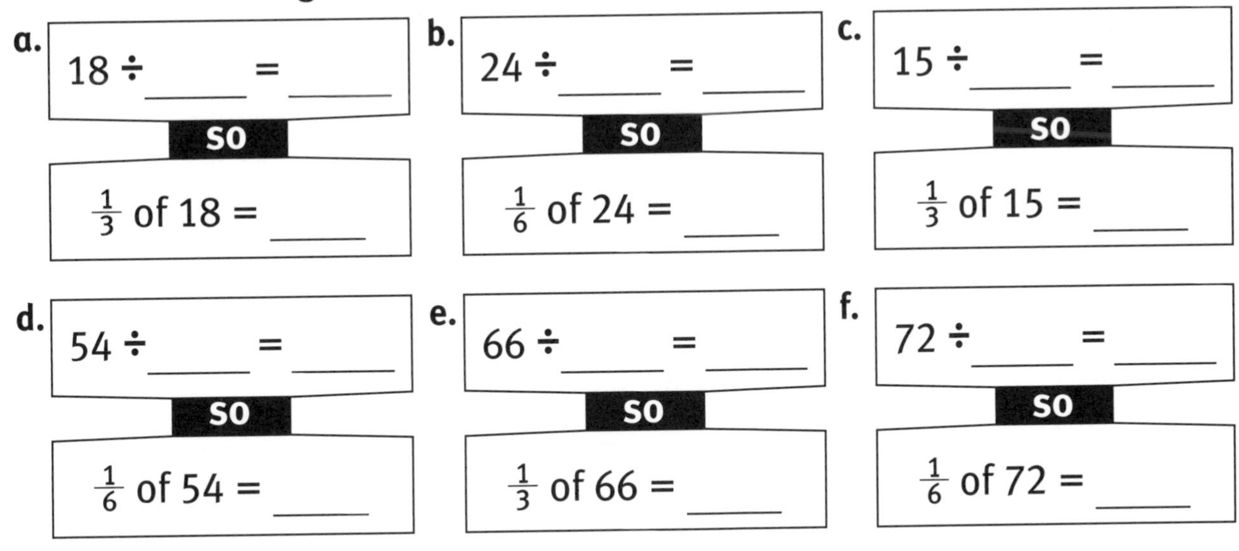

2. a. Write the answers.

$\frac{1}{3}$ of 36 = ____

$\frac{1}{6}$ of 36 = ____

b. Compare the two number sentences above. Write what you notice.

____

____

3. Find the fraction of the numbers in the inner circles.
   Write the answers around the outside.

a.

b.

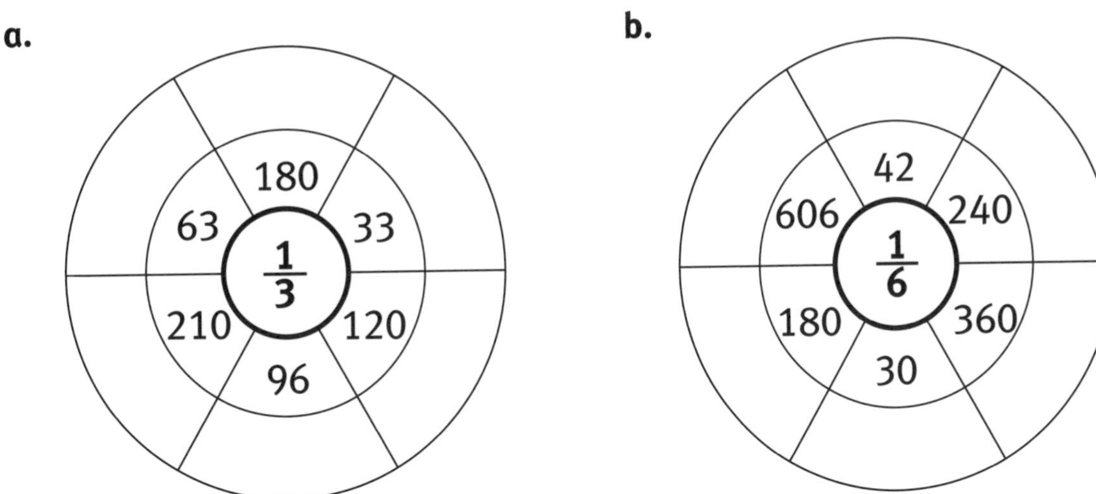

WARM UP 25

Name: ______________________

A farmer used $\frac{1}{10}$ of a new roll of fencing wire.
The roll was 250 yards long.
How many yards did he use?

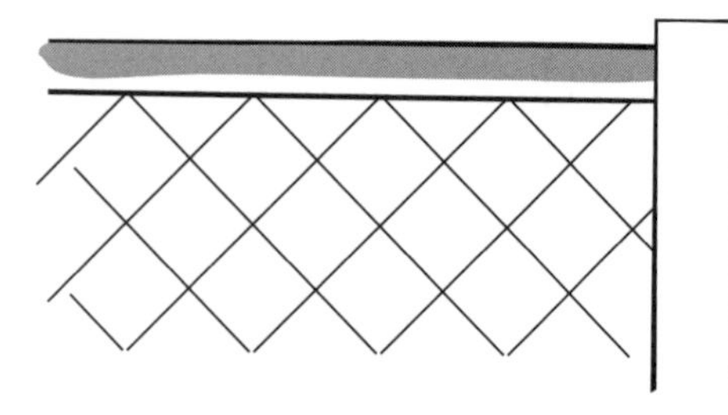

1. Complete this sentence to show one way to calculate the answer.

$\frac{1}{10}$ of 250 **is the same as** 250 ÷ _____ = _____

Describe another way you could figure out $\frac{1}{10}$ of 250?

2. Suppose the farmer used $\frac{1}{5}$ of 250 yards of wire. Complete this sentence to show how much he used.

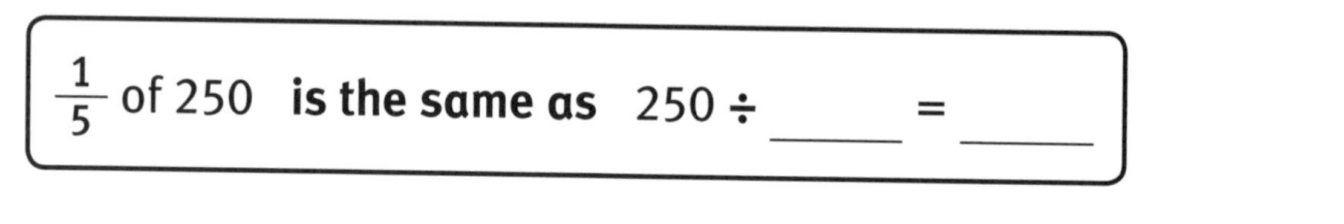
$\frac{1}{5}$ of 250 **is the same as** 250 ÷ _____ = _____

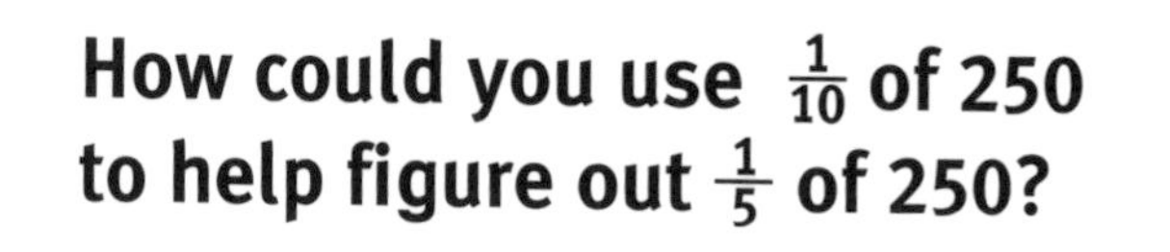

3. Use the same strategy to figure out these in your head.

a. $\frac{1}{10}$ of 180 = _____

b. $\frac{1}{5}$ of 180 = _____

Name: ______________________

1. a. Calculate $\frac{1}{5}$ of 240 in your head. Describe the strategy you used.

____

____

b. Complete two other number sentences you could solve the same way.

$\frac{1}{5}$ of _____ = _____ $\frac{1}{5}$ of _____ = _____

2. Write the missing numbers.

a. $\frac{1}{10}$ of 120 = _____ **SO** $\frac{1}{5}$ of 120 = _____

b. $\frac{1}{10}$ of 230 = _____ **SO** $\frac{1}{5}$ of 230 = _____

c. $\frac{1}{10}$ of 160 = _____ **SO** $\frac{1}{5}$ of 160 = _____

d. $\frac{1}{10}$ of 330 = _____ **SO** $\frac{1}{5}$ of 330 = _____

e. $\frac{1}{10}$ of 290 = _____ **SO** $\frac{1}{5}$ of 290 = _____

f. $\frac{1}{10}$ of 410 = _____ **SO** $\frac{1}{5}$ of 410 = _____

3. Write numbers to make these statements true.

a. $\frac{1}{10}$ of _____ = _____ **SO** $\frac{1}{5}$ of _____ = _____

b. $\frac{1}{10}$ of _____ = _____ **SO** $\frac{1}{5}$ of _____ = _____

c. $\frac{1}{10}$ of _____ = _____ **SO** $\frac{1}{5}$ of _____ = _____

4. Write the answers.

a. $\frac{1}{5}$ of 140 = _____ **b.** $\frac{1}{10}$ of 190 = _____ **c.** $\frac{1}{5}$ of 260 = _____

d. $\frac{1}{10}$ of 720 = _____ **e.** $\frac{1}{5}$ of 440 = _____ **f.** $\frac{1}{10}$ of 380 = _____

CHECK UP 3

Name: ____________________

1. Figure out each of these in your head. Write the answers.

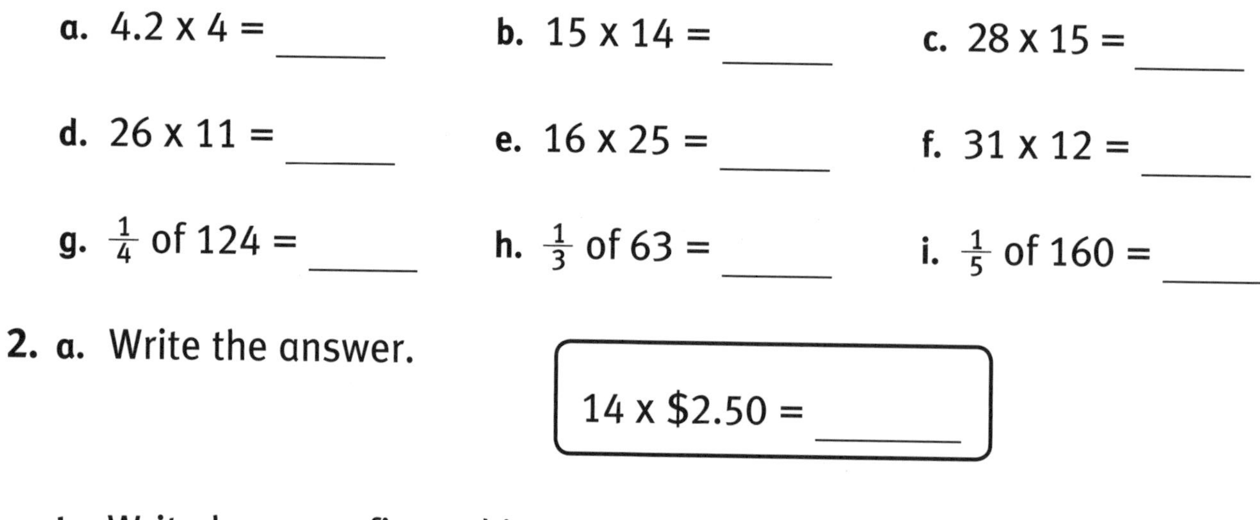

a. $4.2 \times 4 =$ ______ **b.** $15 \times 14 =$ ______ **c.** $28 \times 15 =$ ______

d. $26 \times 11 =$ ______ **e.** $16 \times 25 =$ ______ **f.** $31 \times 12 =$ ______

g. $\frac{1}{4}$ of $124 =$ ______ **h.** $\frac{1}{3}$ of $63 =$ ______ **i.** $\frac{1}{5}$ of $160 =$ ______

2. a. Write the answer.

$14 \times \$2.50 =$ ______

b. Write how you figured it out.

____

____

c. Write some other number sentences you could solve using this strategy.

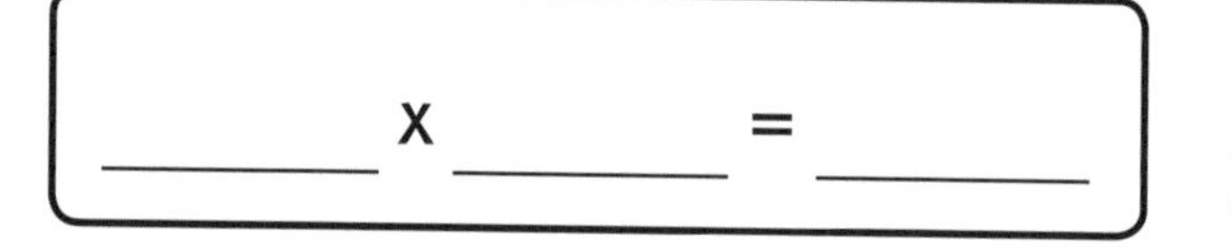

______ x ______ = ______ ______ x ______ = ______

3. For each of these, write the answer then write another number sentence you could solve using the same method.

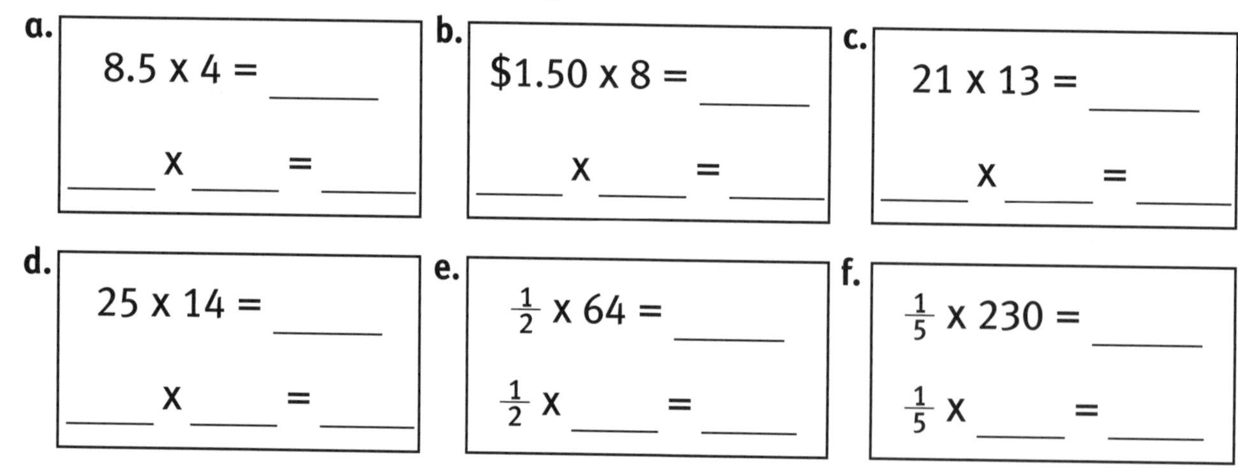

a. $8.5 \times 4 =$ ______

______ x ______ = ______

b. $\$1.50 \times 8 =$ ______

______ x ______ = ______

c. $21 \times 13 =$ ______

______ x ______ = ______

d. $25 \times 14 =$ ______

______ x ______ = ______

e. $\frac{1}{2} \times 64 =$ ______

$\frac{1}{2}$ x ______ = ______

f. $\frac{1}{5} \times 230 =$ ______

$\frac{1}{5}$ x ______ = ______

Name: ______________________

A ship has a ladder with 8 steps.
The water is on the fourth step during a 2-meter high tide.
Which step is the water on during a 1-meter low tide?

Solve this riddle by figuring out the answers below. Write the letters above their matching answers at the bottom of the page. Some letters appear more than once.

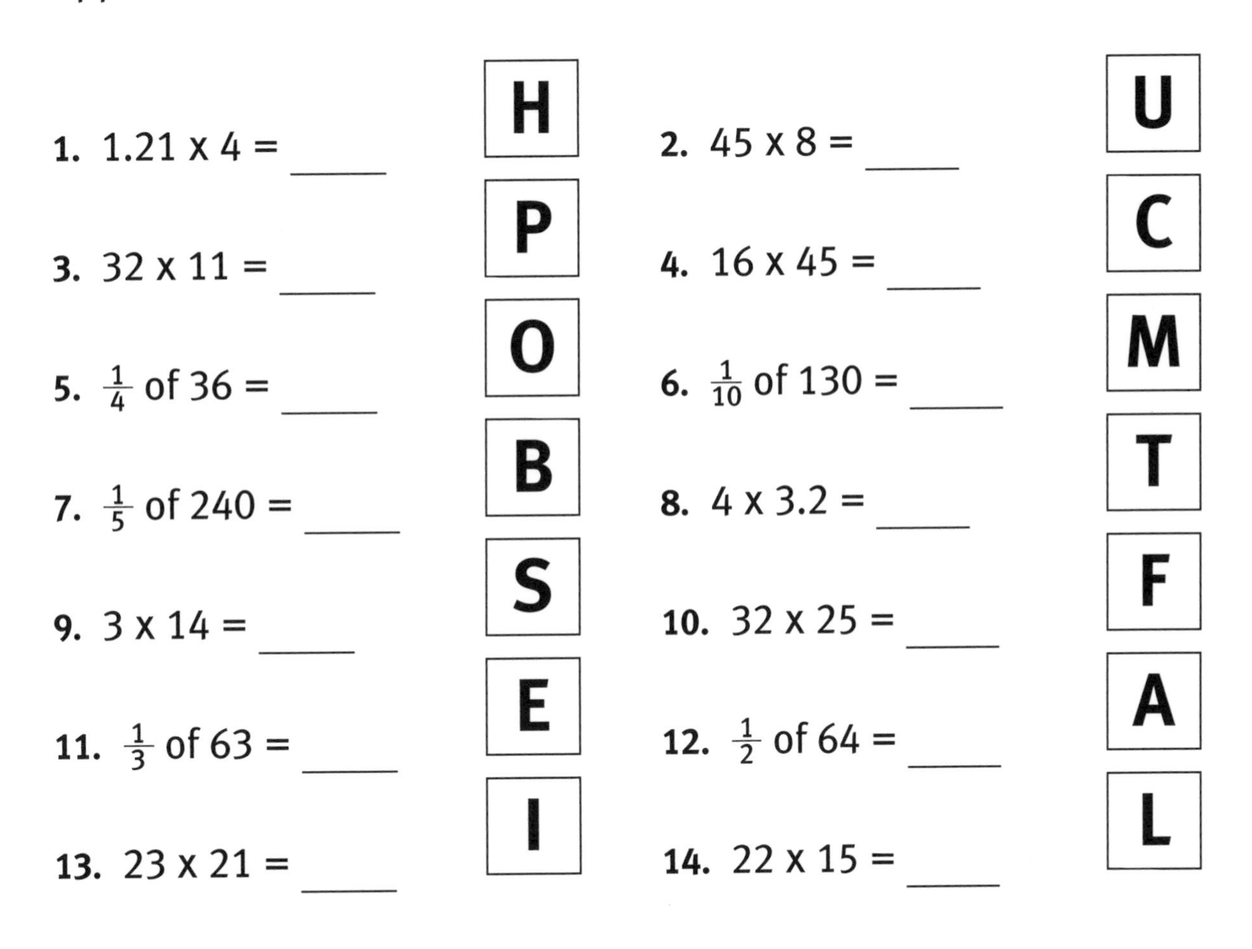

1. 1.21 x 4 = ____ **H**

2. 45 x 8 = ____ **U**

3. 32 x 11 = ____ **P**

4. 16 x 45 = ____ **C**

5. $\frac{1}{4}$ of 36 = ____ **O**

6. $\frac{1}{10}$ of 130 = ____ **M**

7. $\frac{1}{5}$ of 240 = ____ **B**

8. 4 x 3.2 = ____ **T**

9. 3 x 14 = ____ **S**

10. 32 x 25 = ____ **F**

11. $\frac{1}{3}$ of 63 = ____ **E**

12. $\frac{1}{2}$ of 64 = ____ **A**

13. 23 x 21 = ____ **I**

14. 22 x 15 = ____ **L**

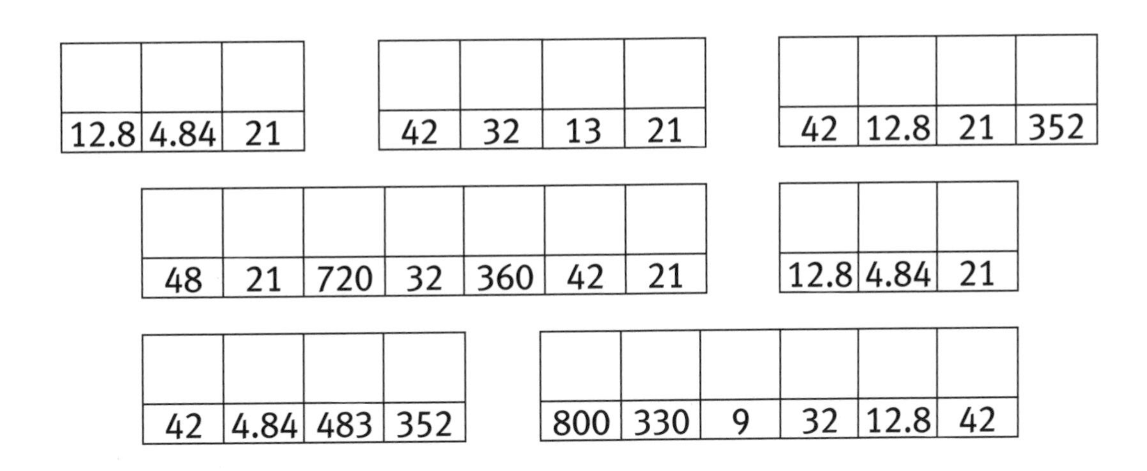

12.8	4.84	21

42	32	13	21

42	12.8	21	352

48	21	720	32	360	42	21

12.8	4.84	21

42	4.84	483	352

800	330	9	32	12.8	42

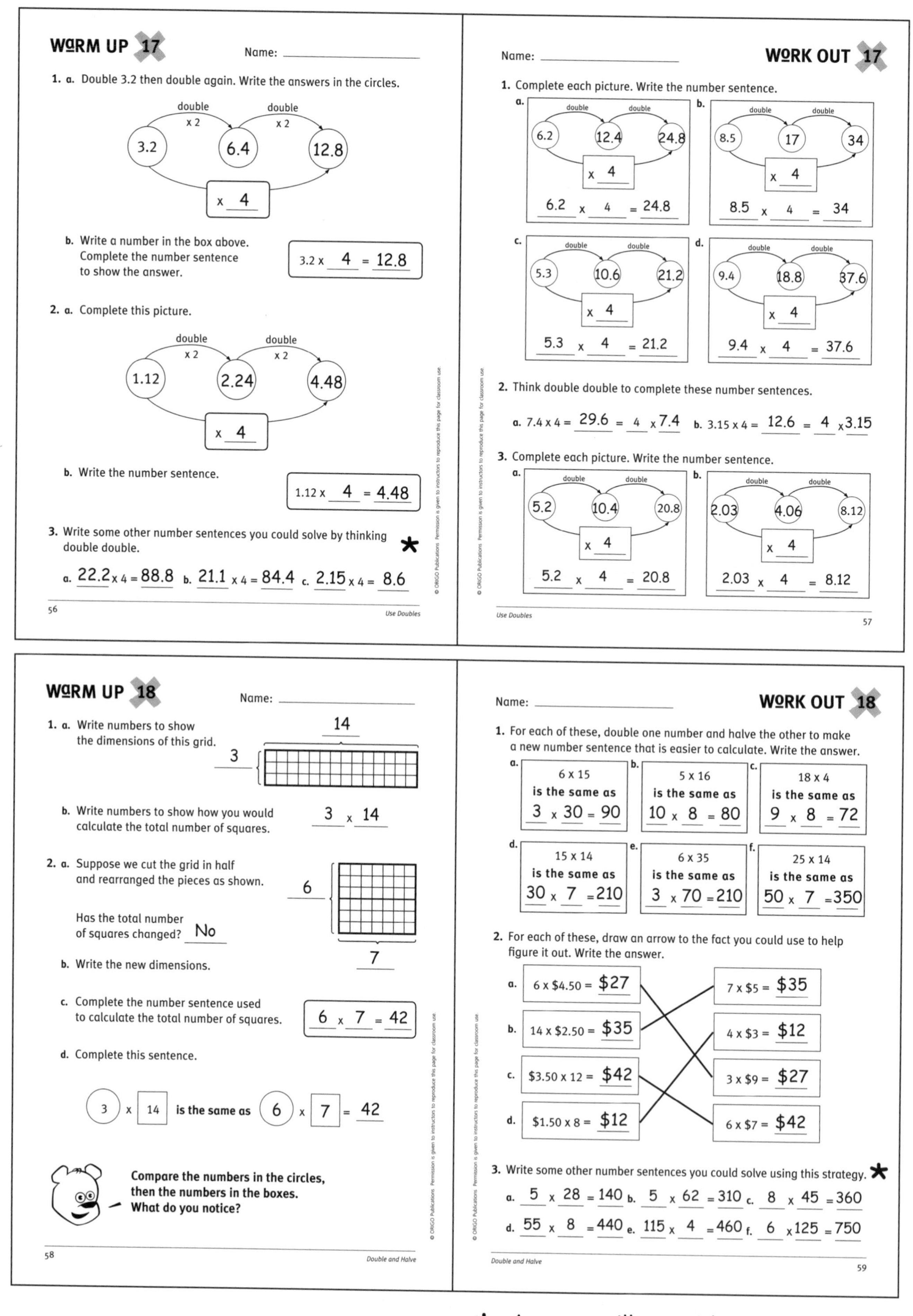

WORM UP 17 Name: ____________

1. a. Double 3.2 then double again. Write the answers in the circles.

double x 2: 3.2 → 6.4; double x 2: 6.4 → 12.8; x 4

b. Write a number in the box above. Complete the number sentence to show the answer.

3.2 x 4 = 12.8

2. a. Complete this picture.

double x 2: 1.12 → 2.24; double x 2: 2.24 → 4.48; x 4

b. Write the number sentence.

1.12 x 4 = 4.48

3. Write some other number sentences you could solve by thinking double double. *

a. 22.2 x 4 = 88.8 b. 21.1 x 4 = 84.4 c. 2.15 x 4 = 8.6

56 *Use Doubles*

© ORIGO Publications Permission is given to instructors to reproduce this page for classroom use.

Name: ____________ **WORK OUT 17**

1. Complete each picture. Write the number sentence.

a. double: 6.2 → 12.4; double: 12.4 → 24.8; x 4
6.2 x 4 = 24.8

b. double: 8.5 → 17; double: 17 → 34; x 4
8.5 x 4 = 34

c. double: 5.3 → 10.6; double: 10.6 → 21.2; x 4
5.3 x 4 = 21.2

d. double: 9.4 → 18.8; double: 18.8 → 37.6; x 4
9.4 x 4 = 37.6

2. Think double double to complete these number sentences.

a. 7.4 x 4 = 29.6 = 4 x 7.4 b. 3.15 x 4 = 12.6 = 4 x 3.15

3. Complete each picture. Write the number sentence.

a. double: 5.2 → 10.4; double: 10.4 → 20.8; x 4
5.2 x 4 = 20.8

b. double: 2.03 → 4.06; double: 4.06 → 8.12; x 4
2.03 x 4 = 8.12

Use Doubles 57

© ORIGO Publications Permission is given to instructors to reproduce this page for classroom use.

WORM UP 18 Name: ____________

1. a. Write numbers to show the dimensions of this grid. 14, 3

b. Write numbers to show how you would calculate the total number of squares. 3 x 14

2. a. Suppose we cut the grid in half and rearranged the pieces as shown. 6, 7

Has the total number of squares changed? No

b. Write the new dimensions.

c. Complete the number sentence used to calculate the total number of squares. 6 x 7 = 42

d. Complete this sentence.

(3) x [14] is the same as (6) x [7] = 42

Compare the numbers in the circles, then the numbers in the boxes. What do you notice?

58 *Double and Halve*

© ORIGO Publications Permission is given to instructors to reproduce this page for classroom use.

Name: ____________ **WORK OUT 18**

1. For each of these, double one number and halve the other to make a new number sentence that is easier to calculate. Write the answer.

a. 6 x 15 is the same as 3 x 30 = 90
b. 5 x 16 is the same as 10 x 8 = 80
c. 18 x 4 is the same as 9 x 8 = 72
d. 15 x 14 is the same as 30 x 7 = 210
e. 6 x 35 is the same as 3 x 70 = 210
f. 25 x 14 is the same as 50 x 7 = 350

2. For each of these, draw an arrow to the fact you could use to help figure it out. Write the answer.

a. 6 x $4.50 = $27 → 3 x $9 = $27
b. 14 x $2.50 = $35 → 7 x $5 = $35
c. $3.50 x 12 = $42 → 6 x $7 = $42
d. $1.50 x 8 = $12 → 4 x $3 = $12

3. Write some other number sentences you could solve using this strategy. *

a. 5 x 28 = 140 b. 5 x 62 = 310 c. 8 x 45 = 360
d. 55 x 8 = 440 e. 115 x 4 = 460 f. 6 x 125 = 750

Double and Halve 59

* Answers will vary. This is one example.

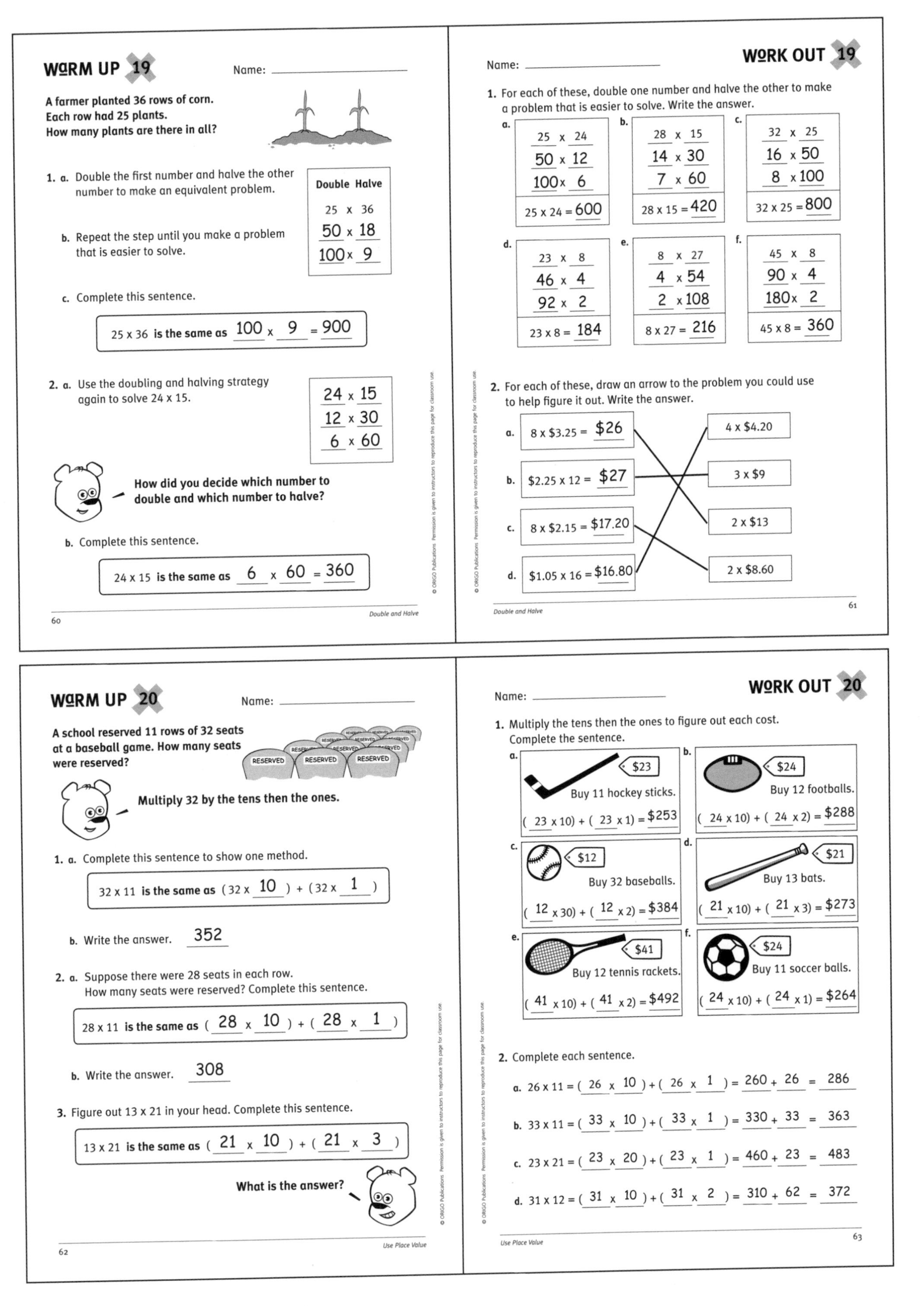

WARM UP 19

Name: ____________

A farmer planted 36 rows of corn. Each row had 25 plants. How many plants are there in all?

1. a. Double the first number and halve the other number to make an equivalent problem.

   b. Repeat the step until you make a problem that is easier to solve.

Double	Halve
25 x	36
50 x	18
100 x	9

   c. Complete this sentence.

   25 x 36 **is the same as** 100 x 9 = 900

2. a. Use the doubling and halving strategy again to solve 24 x 15.

   24 x 15
   12 x 30
   6 x 60

   How did you decide which number to double and which number to halve?

   b. Complete this sentence.

   24 x 15 **is the same as** 6 x 60 = 360

60 *Double and Halve*

WORK OUT 19

Name: ____________

1. For each of these, double one number and halve the other to make a problem that is easier to solve. Write the answer.

   a. 25 x 24 / 50 x 12 / 100 x 6 — 25 x 24 = 600

   b. 28 x 15 / 14 x 30 / 7 x 60 — 28 x 15 = 420

   c. 32 x 25 / 16 x 50 / 8 x 100 — 32 x 25 = 800

   d. 23 x 8 / 46 x 4 / 92 x 2 — 23 x 8 = 184

   e. 8 x 27 / 4 x 54 / 2 x 108 — 8 x 27 = 216

   f. 45 x 8 / 90 x 4 / 180 x 2 — 45 x 8 = 360

2. For each of these, draw an arrow to the problem you could use to help figure it out. Write the answer.

   a. 8 x \$3.25 = \$26 — 2 x \$13
   b. \$2.25 x 12 = \$27 — 3 x \$9
   c. 8 x \$2.15 = \$17.20 — 4 x \$4.20
   d. \$1.05 x 16 = \$16.80 — 2 x \$8.60

Double and Halve 61

WARM UP 20

Name: ____________

A school reserved 11 rows of 32 seats at a baseball game. How many seats were reserved?

RESERVED RESERVED RESERVED

Multiply 32 by the tens then the ones.

1. a. Complete this sentence to show one method.

   32 x 11 **is the same as** (32 x 10) + (32 x 1)

   b. Write the answer. 352

2. a. Suppose there were 28 seats in each row. How many seats were reserved? Complete this sentence.

   28 x 11 **is the same as** (28 x 10) + (28 x 1)

   b. Write the answer. 308

3. Figure out 13 x 21 in your head. Complete this sentence.

   13 x 21 **is the same as** (21 x 10) + (21 x 3)

   What is the answer?

62 *Use Place Value*

WORK OUT 20

Name: ____________

1. Multiply the tens then the ones to figure out each cost. Complete the sentence.

   a. \$23 Buy 11 hockey sticks. (23 x 10) + (23 x 1) = \$253

   b. \$24 Buy 12 footballs. (24 x 10) + (24 x 2) = \$288

   c. \$12 Buy 32 baseballs. (12 x 30) + (12 x 2) = \$384

   d. \$21 Buy 13 bats. (21 x 10) + (21 x 3) = \$273

   e. \$41 Buy 12 tennis rackets. (41 x 10) + (41 x 2) = \$492

   f. \$24 Buy 11 soccer balls. (24 x 10) + (24 x 1) = \$264

2. Complete each sentence.

   a. 26 x 11 = (26 x 10) + (26 x 1) = 260 + 26 = 286

   b. 33 x 11 = (33 x 10) + (33 x 1) = 330 + 33 = 363

   c. 23 x 21 = (23 x 20) + (23 x 1) = 460 + 23 = 483

   d. 31 x 12 = (31 x 10) + (31 x 2) = 310 + 62 = 372

Use Place Value 63

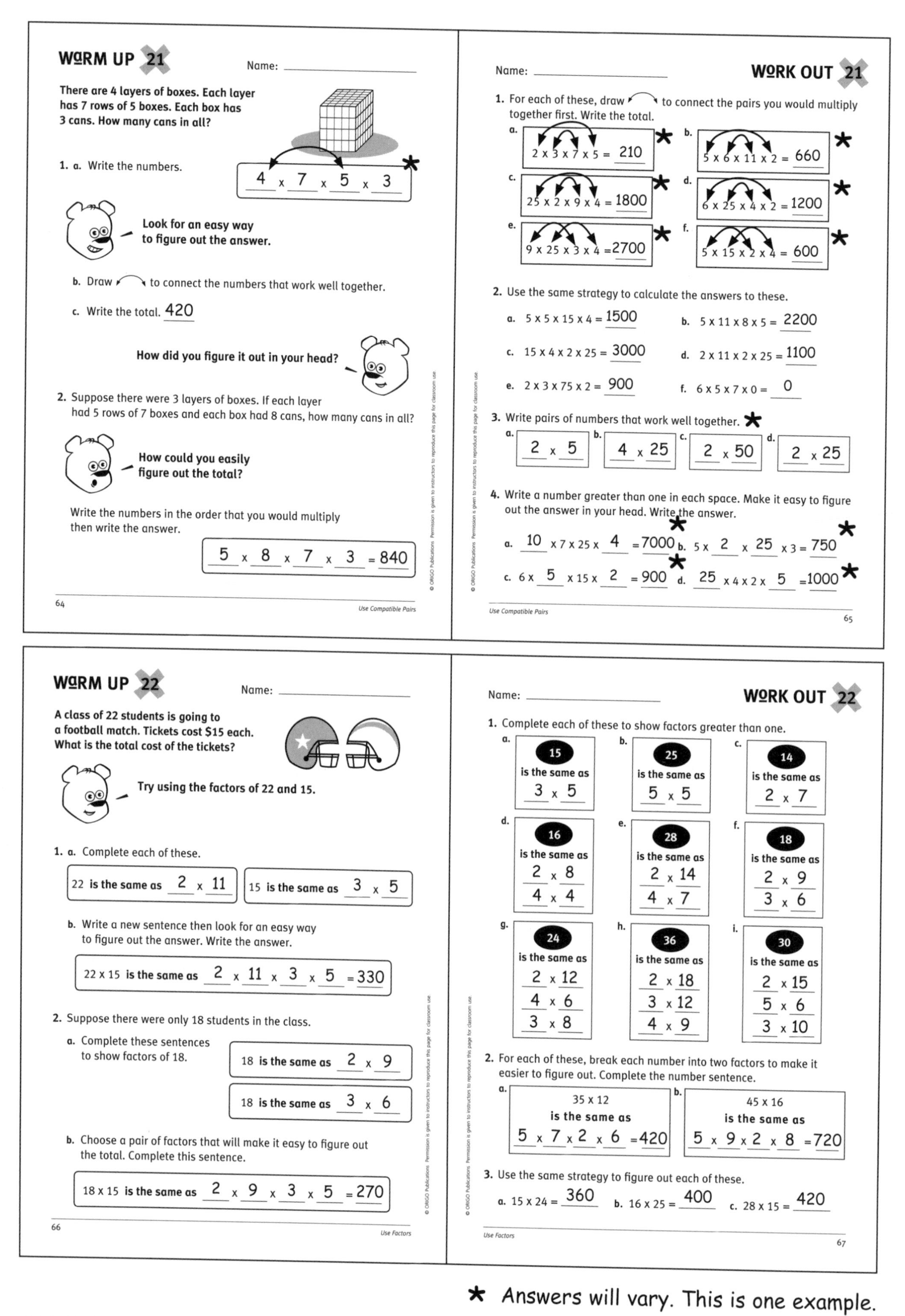

WARM UP 21

Name:

There are 4 layers of boxes. Each layer has 7 rows of 5 boxes. Each box has 3 cans. How many cans in all?

1. a. Write the numbers. 4 x 7 x 5 x 3 *

Look for an easy way to figure out the answer.

b. Draw to connect the numbers that work well together.

c. Write the total. 420

How did you figure it out in your head?

2. Suppose there were 3 layers of boxes. If each layer had 5 rows of 7 boxes and each box had 8 cans, how many cans in all?

How could you easily figure out the total?

Write the numbers in the order that you would multiply then write the answer.

5 x 8 x 7 x 3 = 840

64 Use Compatible Pairs

© ORIGO Publications Permission is given to instructors to reproduce this page for classroom use.

WORK OUT 21

Name:

1. For each of these, draw to connect the pairs you would multiply together first. Write the total.

a. 2 x 3 x 7 x 5 = 210 *
b. 5 x 6 x 11 x 2 = 660 *
c. 25 x 2 x 9 x 4 = 1800 *
d. 6 x 25 x 4 x 2 = 1200 *
e. 9 x 25 x 3 x 4 = 2700 *
f. 5 x 15 x 2 x 4 = 600 *

2. Use the same strategy to calculate the answers to these.

a. 5 x 5 x 15 x 4 = 1500
b. 5 x 11 x 8 x 5 = 2200
c. 15 x 4 x 2 x 25 = 3000
d. 2 x 11 x 2 x 25 = 1100
e. 2 x 3 x 75 x 2 = 900
f. 6 x 5 x 7 x 0 = 0

3. Write pairs of numbers that work well together. *

a. 2 x 5
b. 4 x 25
c. 2 x 50
d. 2 x 25

4. Write a number greater than one in each space. Make it easy to figure out the answer in your head. Write the answer.

a. 10 x 7 x 25 x 4 = 7000 *
b. 5 x 2 x 25 x 3 = 750 *
c. 6 x 5 x 15 x 2 = 900 *
d. 25 x 4 x 2 x 5 = 1000 *

© ORIGO Publications Permission is given to instructors to reproduce this page for classroom use.

Use Compatible Pairs 65

WARM UP 22

Name:

A class of 22 students is going to a football match. Tickets cost $15 each. What is the total cost of the tickets?

Try using the factors of 22 and 15.

1. a. Complete each of these.

22 is the same as 2 x 11

15 is the same as 3 x 5

b. Write a new sentence then look for an easy way to figure out the answer. Write the answer.

22 x 15 is the same as 2 x 11 x 3 x 5 = 330

2. Suppose there were only 18 students in the class.

a. Complete these sentences to show factors of 18.

18 is the same as 2 x 9

18 is the same as 3 x 6

b. Choose a pair of factors that will make it easy to figure out the total. Complete this sentence.

18 x 15 is the same as 2 x 9 x 3 x 5 = 270

66 Use Factors

© ORIGO Publications Permission is given to instructors to reproduce this page for classroom use.

WORK OUT 22

Name:

1. Complete each of these to show factors greater than one.

a. 15 is the same as 3 x 5
b. 25 is the same as 5 x 5
c. 14 is the same as 2 x 7
d. 16 is the same as 2 x 8, 4 x 4
e. 28 is the same as 2 x 14, 4 x 7
f. 18 is the same as 2 x 9, 3 x 6
g. 24 is the same as 2 x 12, 4 x 6, 3 x 8
h. 36 is the same as 2 x 18, 3 x 12, 4 x 9
i. 30 is the same as 2 x 15, 5 x 6, 3 x 10

2. For each of these, break each number into two factors to make it easier to figure out. Complete the number sentence.

a. 35 x 12 is the same as 5 x 7 x 2 x 6 = 420
b. 45 x 16 is the same as 5 x 9 x 2 x 8 = 720

3. Use the same strategy to figure out each of these.

a. 15 x 24 = 360
b. 16 x 25 = 400
c. 28 x 15 = 420

© ORIGO Publications Permission is given to instructors to reproduce this page for classroom use.

Use Factors 67

* Answers will vary. This is one example.

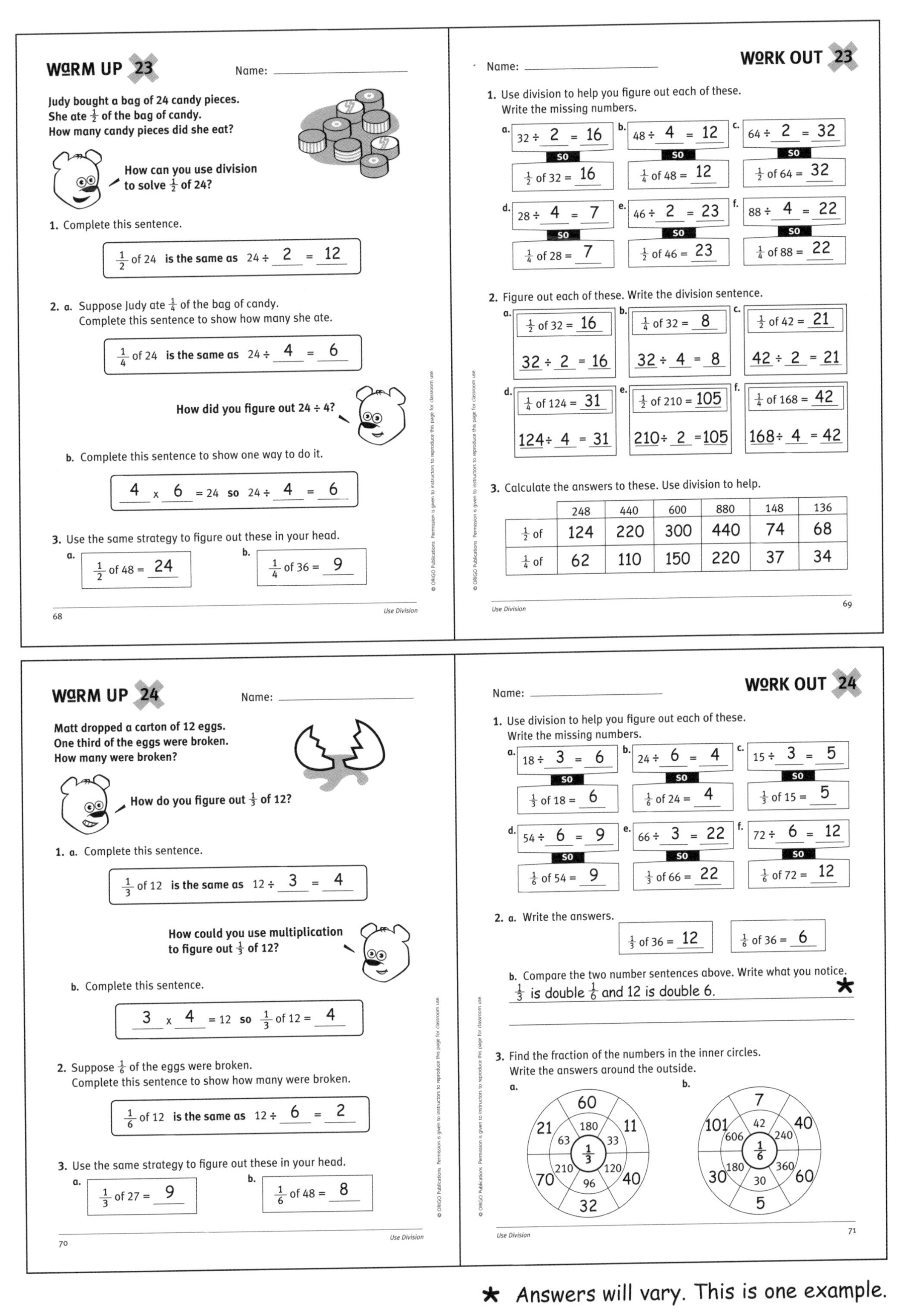

WARM UP 23 Name:

Judy bought a bag of 24 candy pieces. She ate $\frac{1}{2}$ of the bag of candy. How many candy pieces did she eat?

How can you use division to solve $\frac{1}{2}$ of 24?

1. Complete this sentence.

$\frac{1}{2}$ of 24 **is the same as** 24 ÷ 2 = 12

2. a. Suppose Judy ate $\frac{1}{4}$ of the bag of candy. Complete this sentence to show how many she ate.

$\frac{1}{4}$ of 24 **is the same as** 24 ÷ 4 = 6

How did you figure out 24 ÷ 4?

b. Complete this sentence to show one way to do it.

4 x 6 = 24 **so** 24 ÷ 4 = 6

3. Use the same strategy to figure out these in your head.

a. $\frac{1}{2}$ of 48 = 24

b. $\frac{1}{4}$ of 36 = 9

68 *Use Division*

WORK OUT 23 Name:

1. Use division to help you figure out each of these. Write the missing numbers.

a. 32 ÷ 2 = 16 **SO** $\frac{1}{2}$ of 32 = 16

b. 48 ÷ 4 = 12 **SO** $\frac{1}{4}$ of 48 = 12

c. 64 ÷ 2 = 32 **SO** $\frac{1}{2}$ of 64 = 32

d. 28 ÷ 4 = 7 **SO** $\frac{1}{4}$ of 28 = 7

e. 46 ÷ 2 = 23 **SO** $\frac{1}{2}$ of 46 = 23

f. 88 ÷ 4 = 22 **SO** $\frac{1}{4}$ of 88 = 22

2. Figure out each of these. Write the division sentence.

a. $\frac{1}{2}$ of 32 = 16 — 32 ÷ 2 = 16

b. $\frac{1}{4}$ of 32 = 8 — 32 ÷ 4 = 8

c. $\frac{1}{2}$ of 42 = 21 — 42 ÷ 2 = 21

d. $\frac{1}{4}$ of 124 = 31 — 124 ÷ 4 = 31

e. $\frac{1}{2}$ of 210 = 105 — 210 ÷ 2 = 105

f. $\frac{1}{4}$ of 168 = 42 — 168 ÷ 4 = 42

3. Calculate the answers to these. Use division to help.

	248	440	600	880	148	136
$\frac{1}{2}$ of	124	220	300	440	74	68
$\frac{1}{4}$ of	62	110	150	220	37	34

Use Division 69

WARM UP 24 Name:

Matt dropped a carton of 12 eggs. One third of the eggs were broken. How many were broken?

How do you figure out $\frac{1}{3}$ of 12?

1. a. Complete this sentence.

$\frac{1}{3}$ of 12 **is the same as** 12 ÷ 3 = 4

How could you use multiplication to figure out $\frac{1}{3}$ of 12?

b. Complete this sentence.

3 x 4 = 12 **so** $\frac{1}{3}$ of 12 = 4

2. Suppose $\frac{1}{6}$ of the eggs were broken. Complete this sentence to show how many were broken.

$\frac{1}{6}$ of 12 **is the same as** 12 ÷ 6 = 2

3. Use the same strategy to figure out these in your head.

a. $\frac{1}{3}$ of 27 = 9

b. $\frac{1}{6}$ of 48 = 8

70 *Use Division*

WORK OUT 24 Name:

1. Use division to help you figure out each of these. Write the missing numbers.

a. 18 ÷ 3 = 6 **SO** $\frac{1}{3}$ of 18 = 6

b. 24 ÷ 6 = 4 **SO** $\frac{1}{6}$ of 24 = 4

c. 15 ÷ 3 = 5 **SO** $\frac{1}{3}$ of 15 = 5

d. 54 ÷ 6 = 9 **SO** $\frac{1}{6}$ of 54 = 9

e. 66 ÷ 3 = 22 **SO** $\frac{1}{3}$ of 66 = 22

f. 72 ÷ 6 = 12 **SO** $\frac{1}{6}$ of 72 = 12

2. a. Write the answers.

$\frac{1}{3}$ of 36 = 12 $\frac{1}{6}$ of 36 = 6

b. Compare the two number sentences above. Write what you notice. ★

$\frac{1}{3}$ is double $\frac{1}{6}$ and 12 is double 6.

3. Find the fraction of the numbers in the inner circles. Write the answers around the outside.

a. $\frac{1}{3}$: 180 → 60, 33 → 11, 120 → 40, 96 → 32, 210 → 70, 63 → 21

b. $\frac{1}{6}$: 42 → 7, 240 → 40, 360 → 60, 30 → 5, 180 → 30, 606 → 101

Use Division 71

★ Answers will vary. This is one example.

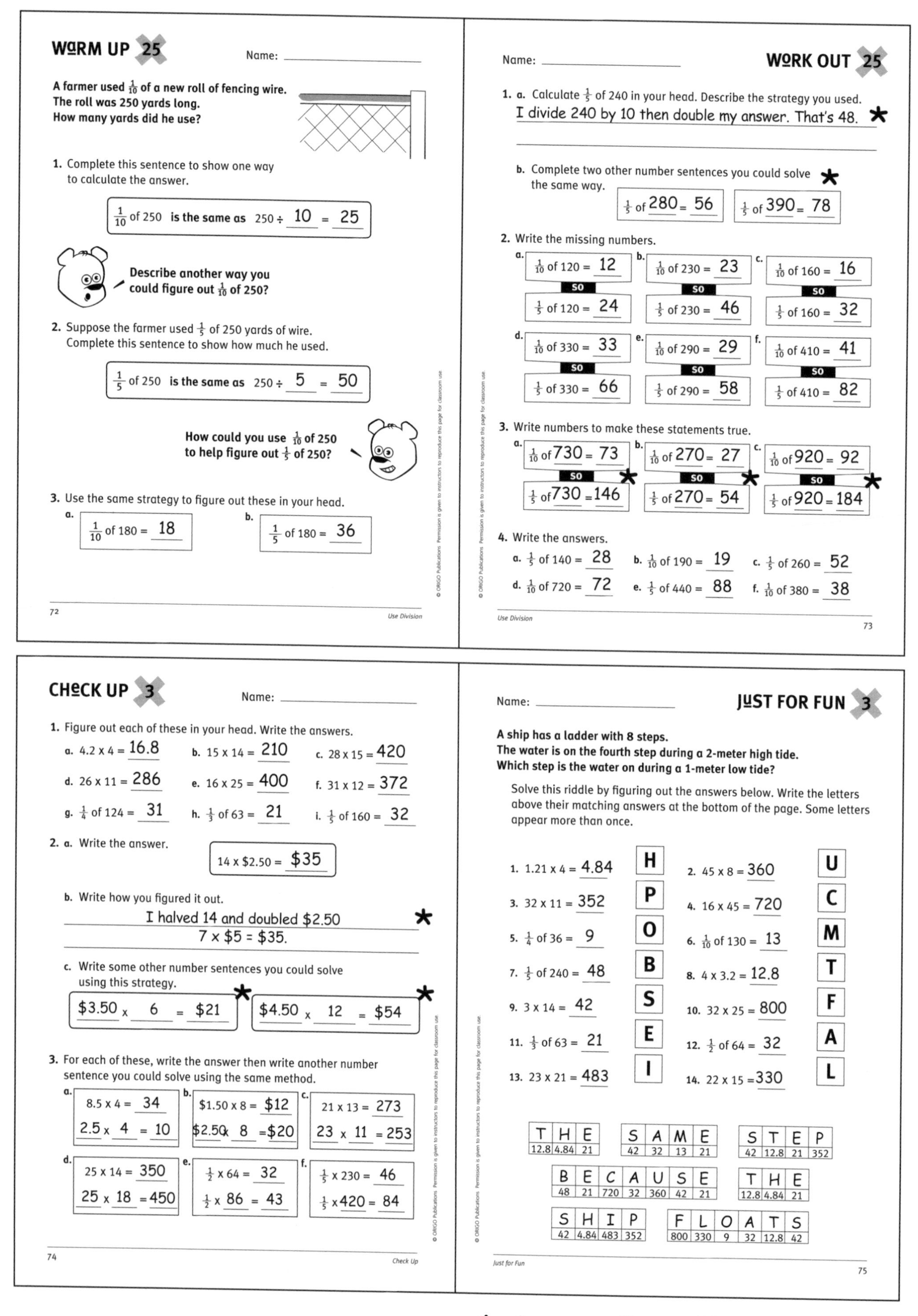

WARM UP 25

Name: ______

A farmer used $\frac{1}{10}$ of a new roll of fencing wire. The roll was 250 yards long. How many yards did he use?

1. Complete this sentence to show one way to calculate the answer.

   $\frac{1}{10}$ of 250 **is the same as** 250 ÷ 10 = 25

Describe another way you could figure out $\frac{1}{10}$ of 250?

2. Suppose the farmer used $\frac{1}{5}$ of 250 yards of wire. Complete this sentence to show how much he used.

   $\frac{1}{5}$ of 250 **is the same as** 250 ÷ 5 = 50

How could you use $\frac{1}{10}$ of 250 to help figure out $\frac{1}{5}$ of 250?

3. Use the same strategy to figure out these in your head.

   a. $\frac{1}{10}$ of 180 = 18
   b. $\frac{1}{5}$ of 180 = 36

72 *Use Division*

WORK OUT 25

Name: ______

1. a. Calculate $\frac{1}{5}$ of 240 in your head. Describe the strategy you used.

   I divide 240 by 10 then double my answer. That's 48. ★

   b. Complete two other number sentences you could solve the same way. ★

   $\frac{1}{5}$ of 280 = 56 $\frac{1}{5}$ of 390 = 78

2. Write the missing numbers.

   a. $\frac{1}{10}$ of 120 = 12 SO $\frac{1}{5}$ of 120 = 24
   b. $\frac{1}{10}$ of 230 = 23 SO $\frac{1}{5}$ of 230 = 46
   c. $\frac{1}{10}$ of 160 = 16 SO $\frac{1}{5}$ of 160 = 32
   d. $\frac{1}{10}$ of 330 = 33 SO $\frac{1}{5}$ of 330 = 66
   e. $\frac{1}{10}$ of 290 = 29 SO $\frac{1}{5}$ of 290 = 58
   f. $\frac{1}{10}$ of 410 = 41 SO $\frac{1}{5}$ of 410 = 82

3. Write numbers to make these statements true.

   a. $\frac{1}{10}$ of 730 = 73 SO $\frac{1}{5}$ of 730 = 146 ★
   b. $\frac{1}{10}$ of 270 = 27 SO $\frac{1}{5}$ of 270 = 54 ★
   c. $\frac{1}{10}$ of 920 = 92 SO $\frac{1}{5}$ of 920 = 184 ★

4. Write the answers.

   a. $\frac{1}{5}$ of 140 = 28
   b. $\frac{1}{10}$ of 190 = 19
   c. $\frac{1}{5}$ of 260 = 52
   d. $\frac{1}{10}$ of 720 = 72
   e. $\frac{1}{5}$ of 440 = 88
   f. $\frac{1}{10}$ of 380 = 38

Use Division 73

CHECK UP 3

Name: ______

1. Figure out each of these in your head. Write the answers.

   a. 4.2 x 4 = 16.8
   b. 15 x 14 = 210
   c. 28 x 15 = 420
   d. 26 x 11 = 286
   e. 16 x 25 = 400
   f. 31 x 12 = 372
   g. $\frac{1}{4}$ of 124 = 31
   h. $\frac{1}{3}$ of 63 = 21
   i. $\frac{1}{5}$ of 160 = 32

2. a. Write the answer. 14 x \$2.50 = \$35

   b. Write how you figured it out.

   I halved 14 and doubled \$2.50 ★
   7 x \$5 = \$35.

   c. Write some other number sentences you could solve using this strategy.

   \$3.50 x 6 = \$21 ★ \$4.50 x 12 = \$54 ★

3. For each of these, write the answer then write another number sentence you could solve using the same method.

   a. 8.5 x 4 = 34; 2.5 x 4 = 10
   b. \$1.50 x 8 = \$12; \$2.50 x 8 = \$20
   c. 21 x 13 = 273; 23 x 11 = 253
   d. 25 x 14 = 350; 25 x 18 = 450
   e. $\frac{1}{2}$ x 64 = 32; $\frac{1}{2}$ x 86 = 43
   f. $\frac{1}{5}$ x 230 = 46; $\frac{1}{5}$ x 420 = 84

74 *Check Up*

JUST FOR FUN 3

Name: ______

A ship has a ladder with 8 steps. The water is on the fourth step during a 2-meter high tide. Which step is the water on during a 1-meter low tide?

Solve this riddle by figuring out the answers below. Write the letters above their matching answers at the bottom of the page. Some letters appear more than once.

1. 1.21 x 4 = 4.84 **H**
2. 45 x 8 = 360 **U**
3. 32 x 11 = 352 **P**
4. 16 x 45 = 720 **C**
5. $\frac{1}{4}$ of 36 = 9 **O**
6. $\frac{1}{10}$ of 130 = 13 **M**
7. $\frac{1}{5}$ of 240 = 48 **B**
8. 4 x 3.2 = 12.8 **T**
9. 3 x 14 = 42 **S**
10. 32 x 25 = 800 **F**
11. $\frac{1}{3}$ of 63 = 21 **E**
12. $\frac{1}{2}$ of 64 = 32 **A**
13. 23 x 21 = 483 **I**
14. 22 x 15 = 330 **L**

T	H	E
12.8	4.84	21

S	A	M	E
42	32	13	21

S	T	E	P
42	12.8	21	352

B	E	C	A	U	S	E
48	21	720	32	360	42	21

T	H	E
12.8	4.84	21

S	H	I	P
42	4.84	483	352

F	L	O	A	T	S
800	330	9	32	12.8	42

Just for Fun 75

★ Answers will vary. This is one example.

DIVISION STRATEGIES

Halve

124 ÷ 4 is the same as (124 ÷ 2) ÷ 2

Divide the parts

728 ÷ 7 is the same as (700 ÷ 7) + (28 ÷ 7)

$35.50 ÷ 5 is the same as ($35 ÷ 5) + (50¢ ÷ 5)

Break up the dividend

138 ÷ 3 is the same as (120 ÷ 3) + (18 ÷ 3)

Round or adjust

315 ÷ 5 is the same as 630 ÷ 10

350 ÷ 50 is the same as 700 ÷ 100

225 ÷ 25 is the same as 900 ÷ 100

Name: ______________________

Four friends shared the cost of a dinner. The total cost was $124. How much was each person's share?

1. a. Halve 124 then halve again. Write the answers in the circles.

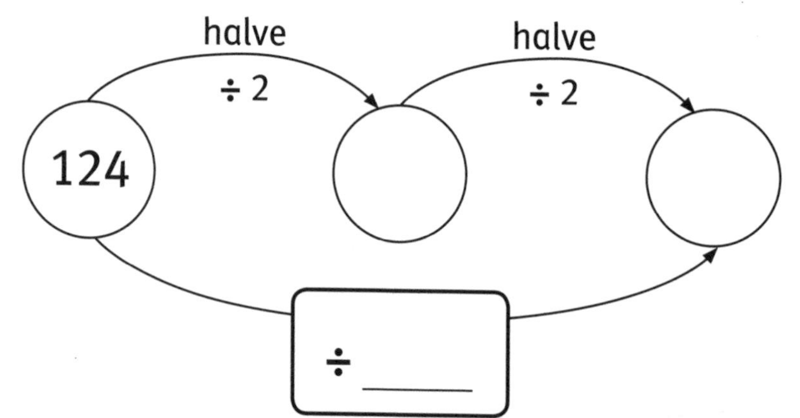

b. Write a number in the box above.
Complete this number sentence.

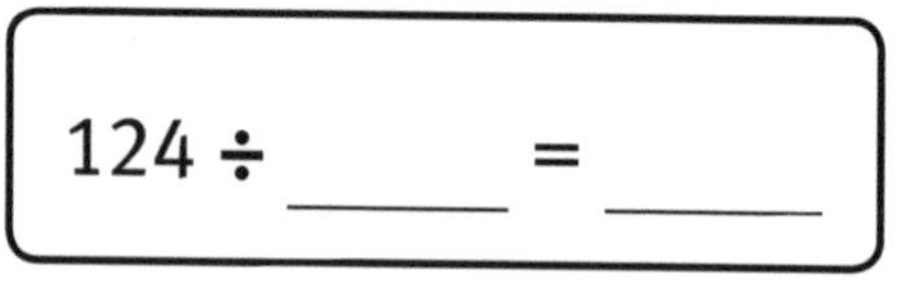

2. a. Complete this picture.

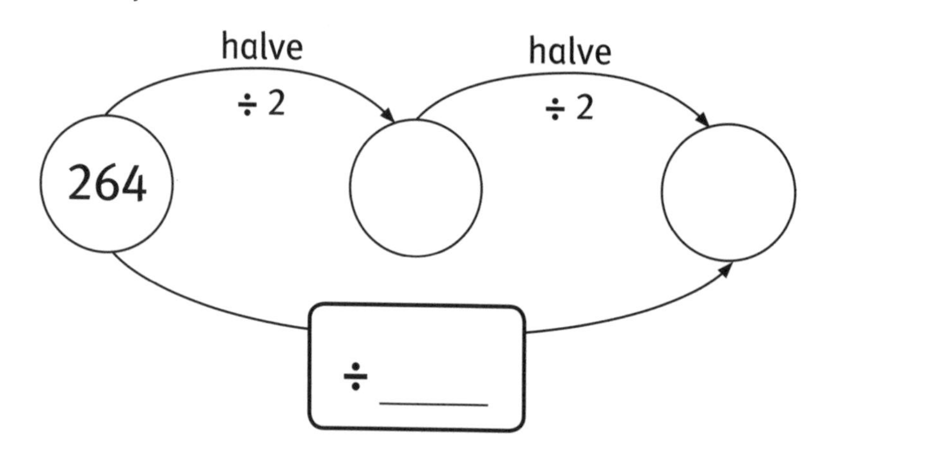

b. Complete this number sentence.

264 ÷ ____ = ____

3. Write some number sentences you can solve by thinking halve halve.

____ ÷ 4 = ____

____ ÷ 4 = ____

____ ÷ 4 = ____

Name: ______________________

1. Complete each picture. Write the number sentence.

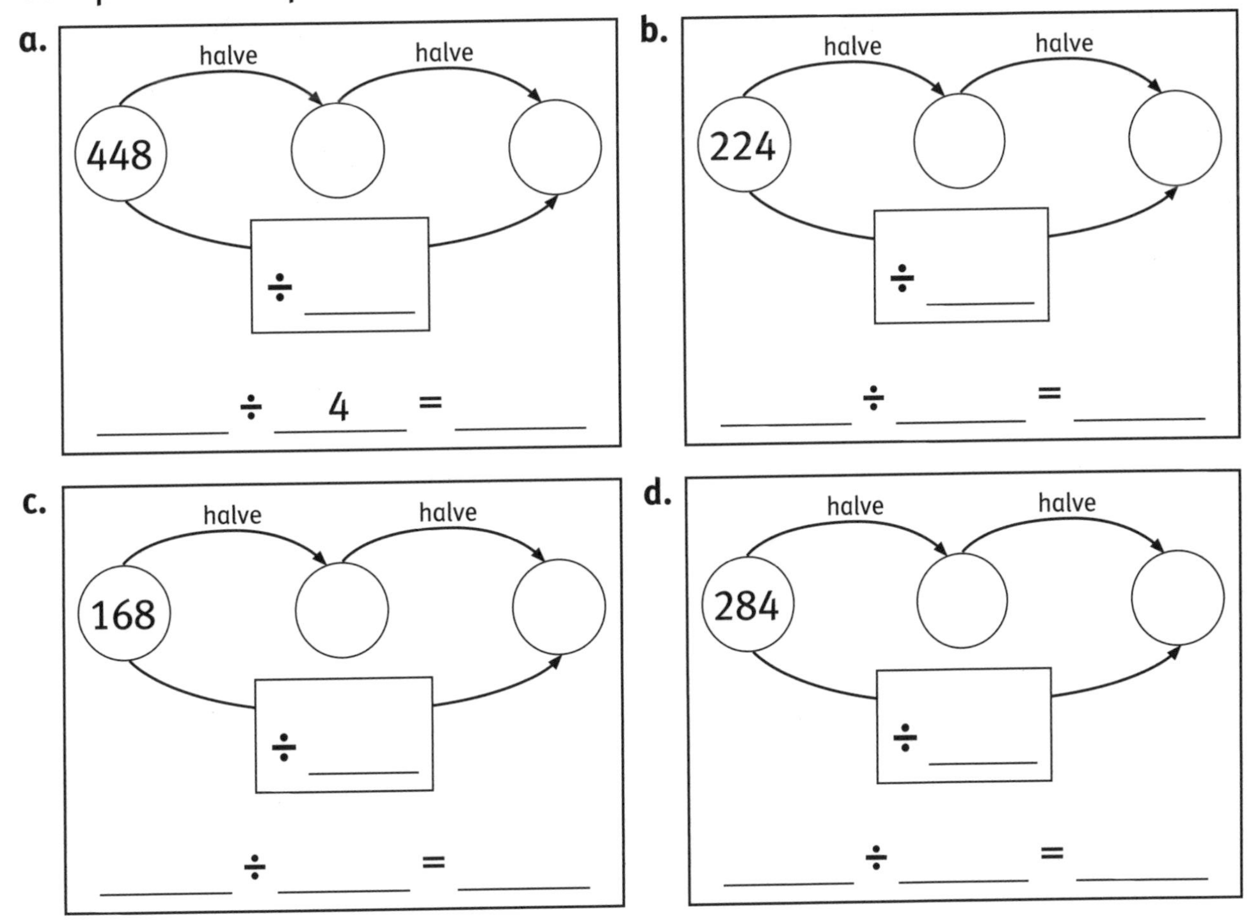

2. For each of these, halve the number then halve again.
Write the number sentence.

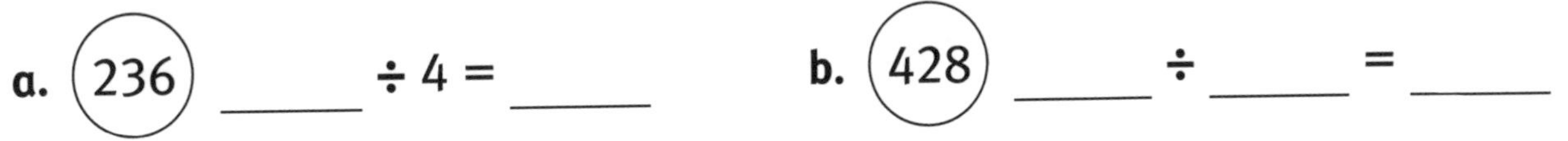

3. Complete each picture. Write the number sentence.

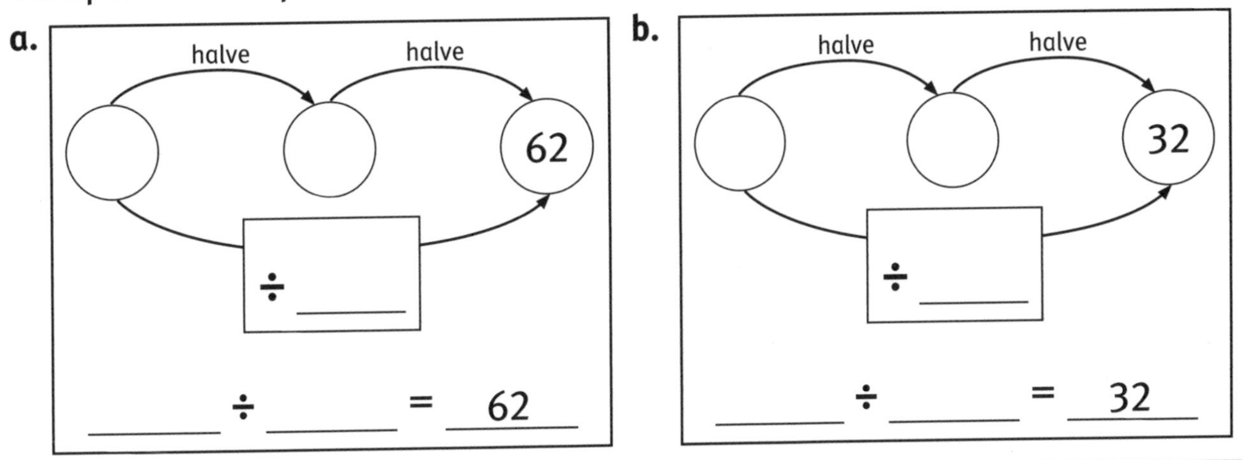

Name: ______________________

A train is moving 728 tons of coal.
If each of the 7 cars carry the same amount, how many tons does each hold?

Try expanding 728 to figure this out in your head.

1. a. Complete this sentence.

728 ÷ 7 **is the same as** 700 ÷ 7 plus ______ ÷ 7

b. Write the answer.

728 ÷ 7 = ______

2. Suppose there were 636 tons of coal and 6 cars.

Try expanding 636 to figure out how much each car would hold.

Complete this sentence.

636 ÷ 6 **is the same as** ______ ÷ 6 plus ______ ÷ 6

What is another way you could expand 636?

Name: ______________________

1. Look at this number sentence.
Write how you know it is correct.

832 ÷ 8 = 104

____

____

2. Complete each sentence. Write the answer.

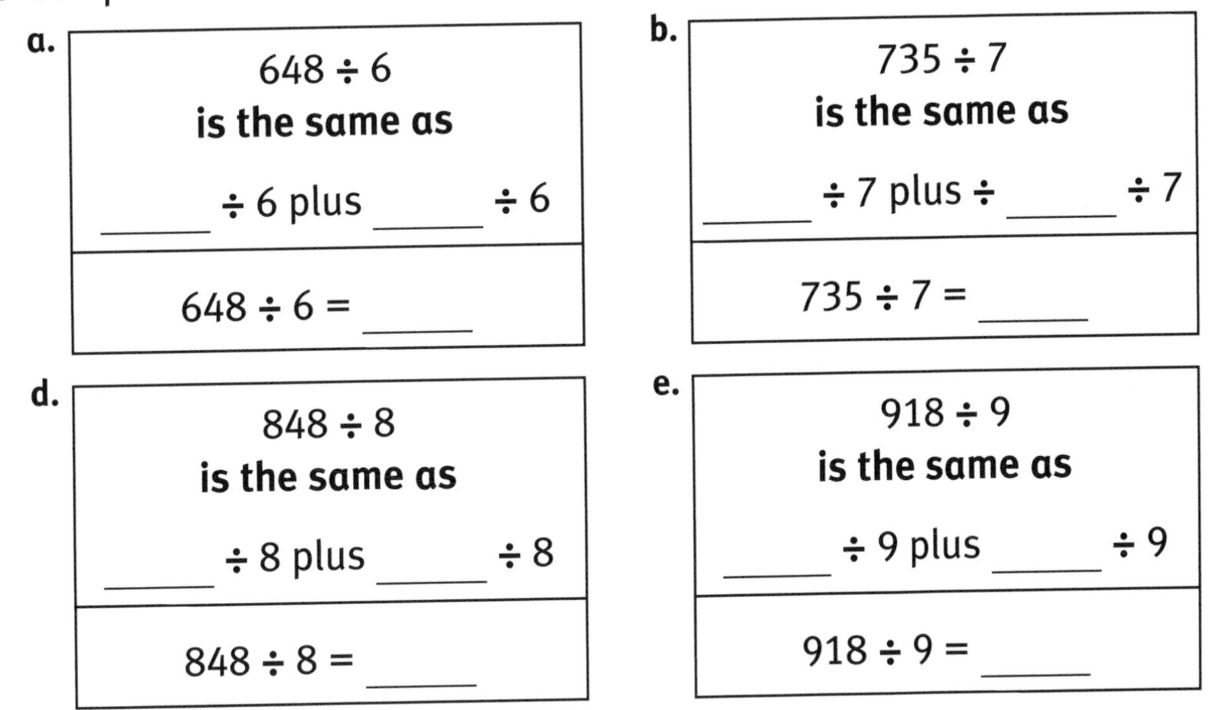

a.
648 ÷ 6
is the same as
______ ÷ 6 plus ______ ÷ 6

648 ÷ 6 = ______

b.
735 ÷ 7
is the same as
______ ÷ 7 plus ÷ ______ ÷ 7

735 ÷ 7 = ______

d.
848 ÷ 8
is the same as
______ ÷ 8 plus ______ ÷ 8

848 ÷ 8 = ______

e.
918 ÷ 9
is the same as
______ ÷ 9 plus ______ ÷ 9

918 ÷ 9 = ______

3. For each of these, draw an arrow to the matching number sentence.
Write the answer.

a. 600 ÷ 6 plus 18 ÷ 6

b. 700 ÷ 7 plus 21 ÷ 7

c. 800 ÷ 8 plus 72 ÷ 8

d. 900 ÷ 9 plus 27 ÷ 9

721 ÷ 7 = ______

872 ÷ 8 = ______

618 ÷ 6 = ______

927 ÷ 9 = ______

Name: ____________________

A 5-day pass costs $35.50.
What is the cost of one day's travel?

Try dividing the dollars and cents separately.

1. a. Complete this sentence.

$35.50 ÷ 5 **is the same as** $______ ÷ 5 plus ______¢ ÷ 5

b. Write the answer. ______

2. Suppose the pass was $25.15.

Figure out the cost of one day's travel by dividing the dollars then the cents.

a. Complete this sentence.

$25.15 ÷ 5 **is the same as** $______ ÷ 5 plus ______¢ ÷ 5

b. Write the answer. ______

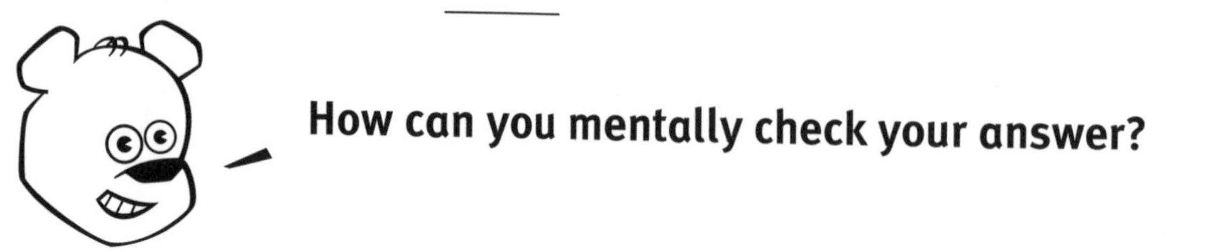

Name: ____________________

1. Lenny figured that $24.32 ÷ 4 is $6.80.
What did he do wrong?

____

____

2. For each of these, complete the sentence then write the answer.

a. $36.48 ÷ 4 **is the same as** $____ ÷ 4 plus ____¢ ÷ 4 = ______

b. $32.40 ÷ 8 **is the same as** $____ ÷ 8 plus ____¢ ÷ 8 = ______

c. $28.56 ÷ 7 **is the same as** $____ ÷ 7 plus ____¢ ÷ 7 = ______

d. $20.32 ÷ 4 **is the same as** $____ ÷ 4 plus ____¢ ÷ 4 = ______

3. Calculate the answers to these. Write number sentences.

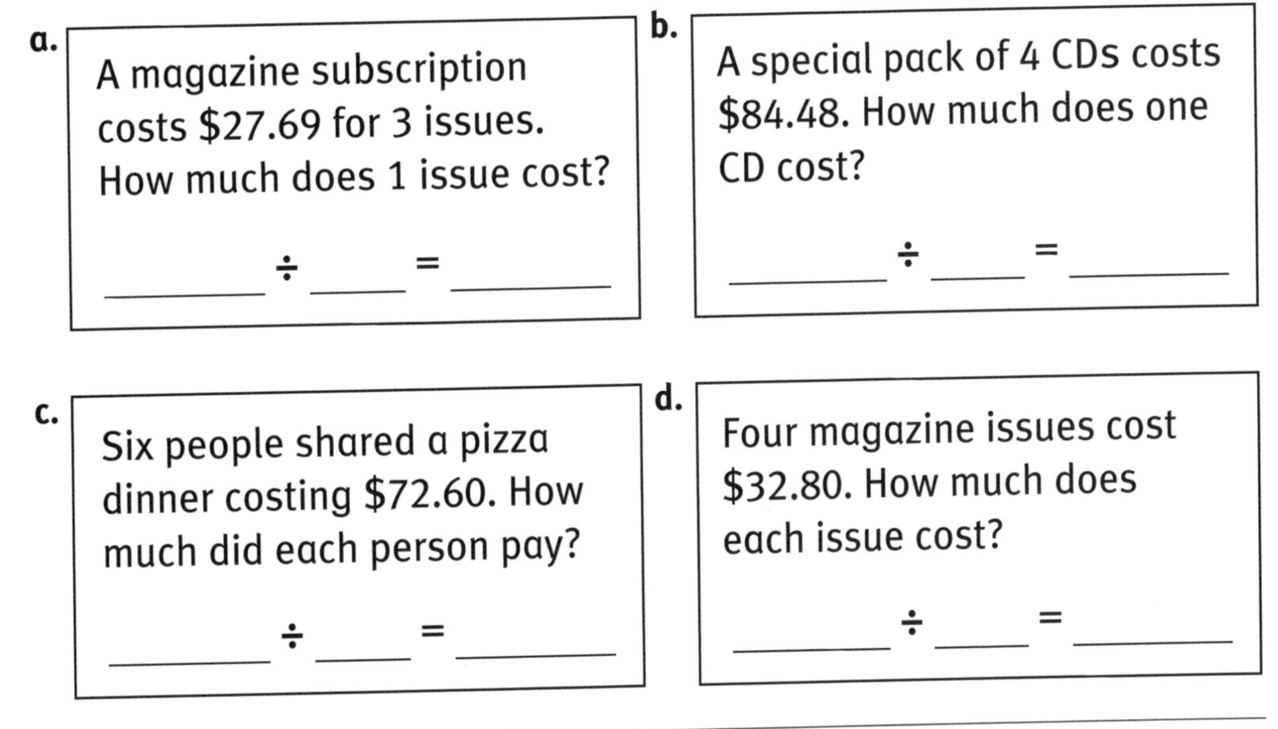

a. A magazine subscription costs $27.69 for 3 issues. How much does 1 issue cost?

______ ÷ ____ = ______

b. A special pack of 4 CDs costs $84.48. How much does one CD cost?

______ ÷ ____ = ______

c. Six people shared a pizza dinner costing $72.60. How much did each person pay?

______ ÷ ____ = ______

d. Four magazine issues cost $32.80. How much does each issue cost?

______ ÷ ____ = ______

WaRM UP 29

Name: ____________________

A school bought 3 new baseball helmets. The total bill was $138. How much did each helmet cost?

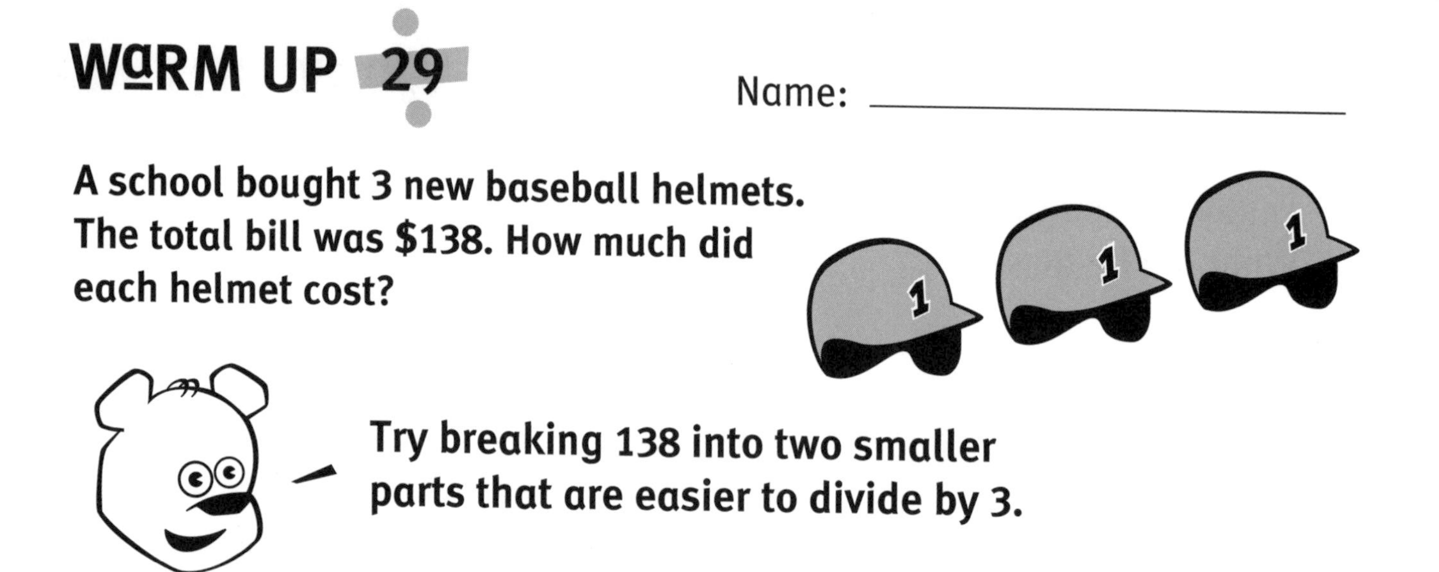

Try breaking 138 into two smaller parts that are easier to divide by 3.

1. a. Complete this sentence to show your thinking.

138 ÷ 3 **is the same as** ______ ÷ 3 plus ______ ÷ 3

b. Write the answer. ______

2. Suppose the total bill was only $117.
Figure out the cost of each helmet by breaking 117 into smaller parts.

a. Complete this sentence.

117 ÷ 3 **is the same as** ______ ÷ 3 plus ______ ÷ 3

b. Write the answer. ______

What are some other ways you could calculate 138 ÷ 3 and 117 ÷ 3 in your head?

Name: ____________________

1. Suppose you had to divide each of these by 4.
   Break each number into two parts that are easier to divide by 4.

   a. 136 → ☐ + ☐
   b. 156 → ☐ + ☐
   c. 176 → ☐ + ☐
   d. 260 → ☐ + ☐

2. Break each of these into parts that you can easily divide by 6.

   a. 156 → ☐ + ☐
   b. 492 → ☐ + ☐
   c. 450 → ☐ + ☐
   d. 216 → ☐ + ☐

3. Figure out the answers to these by breaking the numbers into two smaller parts. Write the answers.

   a. 228 ÷ 3 **is the same as** ______ ÷ 3 plus ______ ÷ 3 = ______

   b. 344 ÷ 8 **is the same as** ______ ÷ 8 plus ______ ÷ 8 = ______

   c. 297 ÷ 9 **is the same as** ______ ÷ 9 plus ______ ÷ 9 = ______

   d. 161 ÷ 7 **is the same as** ______ ÷ 7 plus ______ ÷ 7 = ______

4. Use the same strategy to help you figure out each of these.
   Write the answer.

   a. 198 ÷ 9 = ______
   b. 224 ÷ 7 = ______
   c. 112 ÷ 8 = ______
   d. 115 ÷ 5 = ______
   e. 129 ÷ 3 = ______
   f. 272 ÷ 8 = ______

WARM UP 30

Name: ____________________

Five friends shared the cost of renting a car. The total cost was $315. How much did they each pay?

1. a. Write the division problem.

_______ ÷ _______

Try doubling each number to make an equivalent problem.

b. Complete this sentence.

315 ÷ 5 **is the same as** _______ ÷ _______ = _______

Use a different strategy to check your answer.

2. Adjust 450 ÷ 50 to make an equivalent problem that is easier to solve. Complete this sentence.

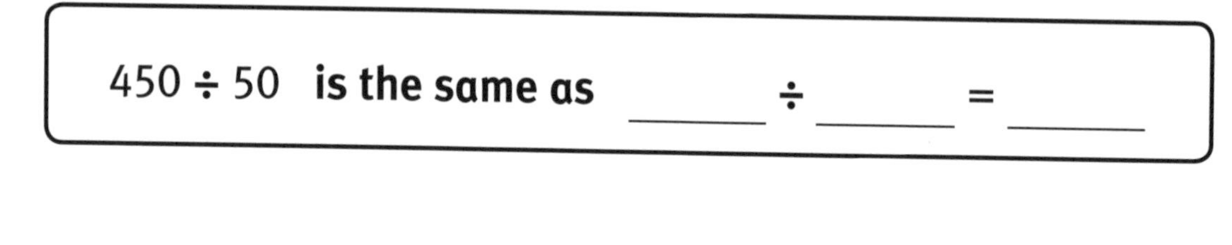

450 ÷ 50 **is the same as** _______ ÷ _______ = _______

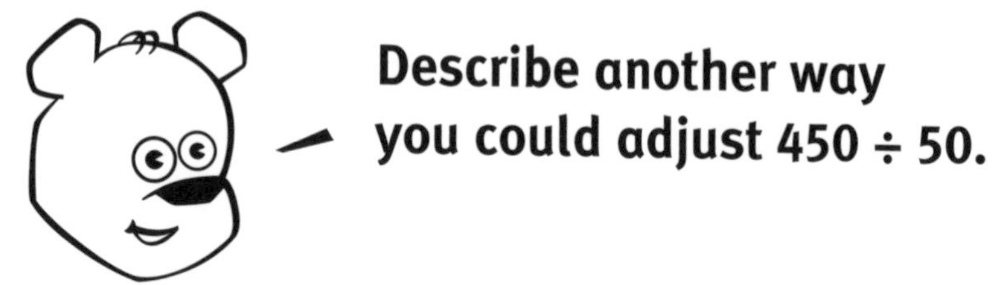

WORK OUT 30

Name: ______________________

1. For each number sentence, double both numbers to make a new problem that is easier to figure out. Write the answer.

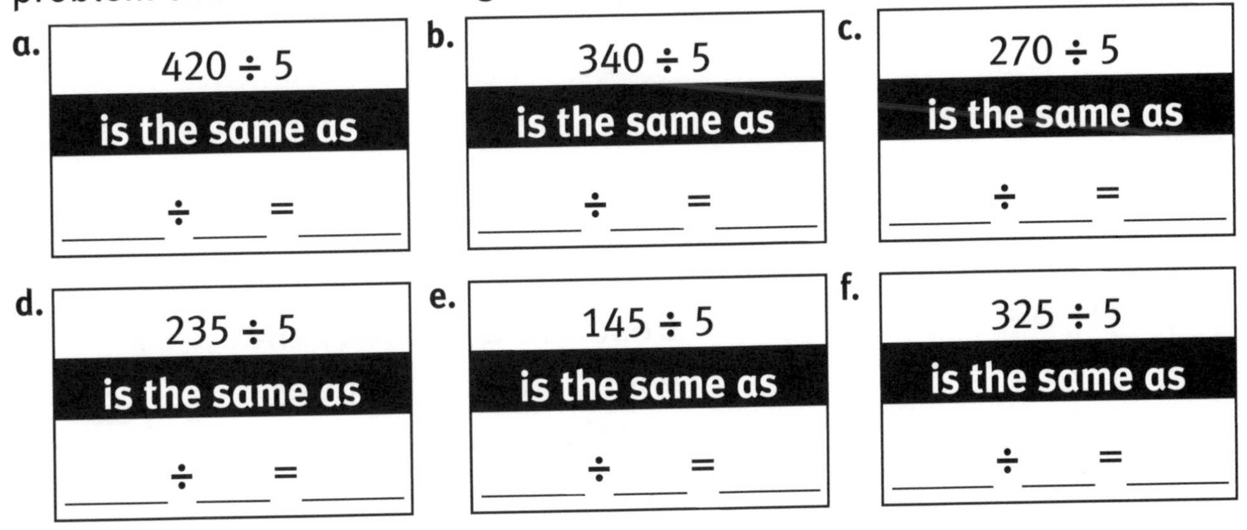

2. Adjust these to make new number sentences that are easier to figure out. There is more than one way to do this.

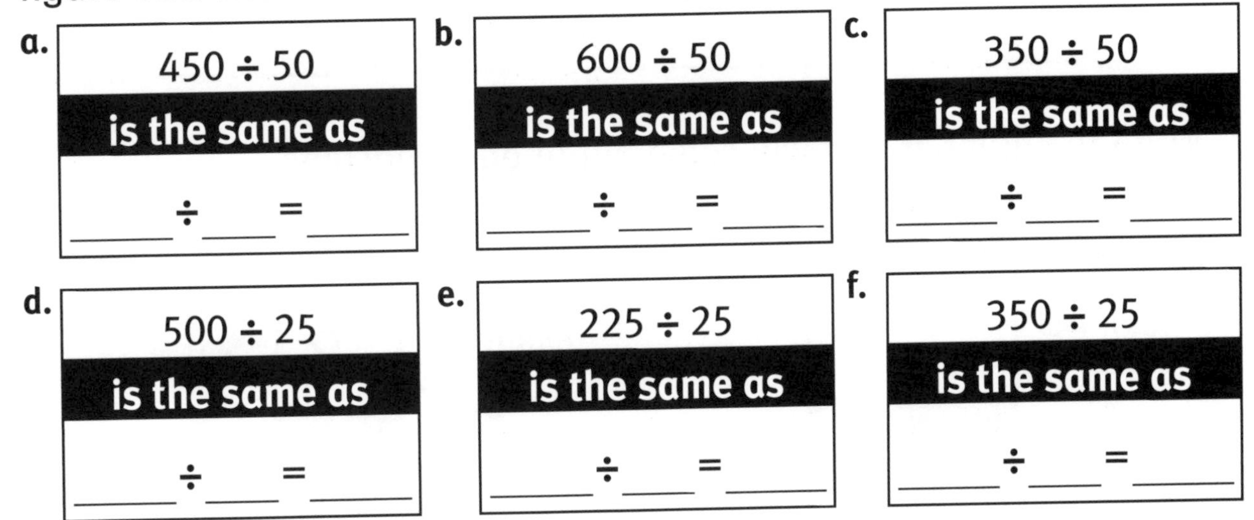

3. Use the same strategy to calculate the answers to these.

a. $12.40 ÷ 5 = ________

b. $21.50 ÷ 5 = ________

c. $12.50 ÷ 50 = ________

d. $32.50 ÷ 50 = ________

e. $6.25 ÷ 25 = ________

f. $12.25 ÷ 25 = ________

CHECK UP 4

Name: ______________________

1. Figure out each of these in your head. Write the answer.

a. 284 ÷ 4 = ______ **b.** 618 ÷ 6 = ______ **c.** 742 ÷ 7 = ______

d. \$15.35 ÷ 5 = ______ **e.** 135 ÷ 3 = ______ **f.** 235 ÷ 5 = ______

2. Look at this number sentence.

\$11.50 ÷ 5 = ______

Describe a strategy that would help you figure out the answer.

a. Write the answer above.

b. Write three other number sentences you can solve the same way.

______ ÷ ____ = ______ | ______ ÷ ____ = ______ | ______ ÷ ____ = ______

3. Write the number that will come out of each machine.

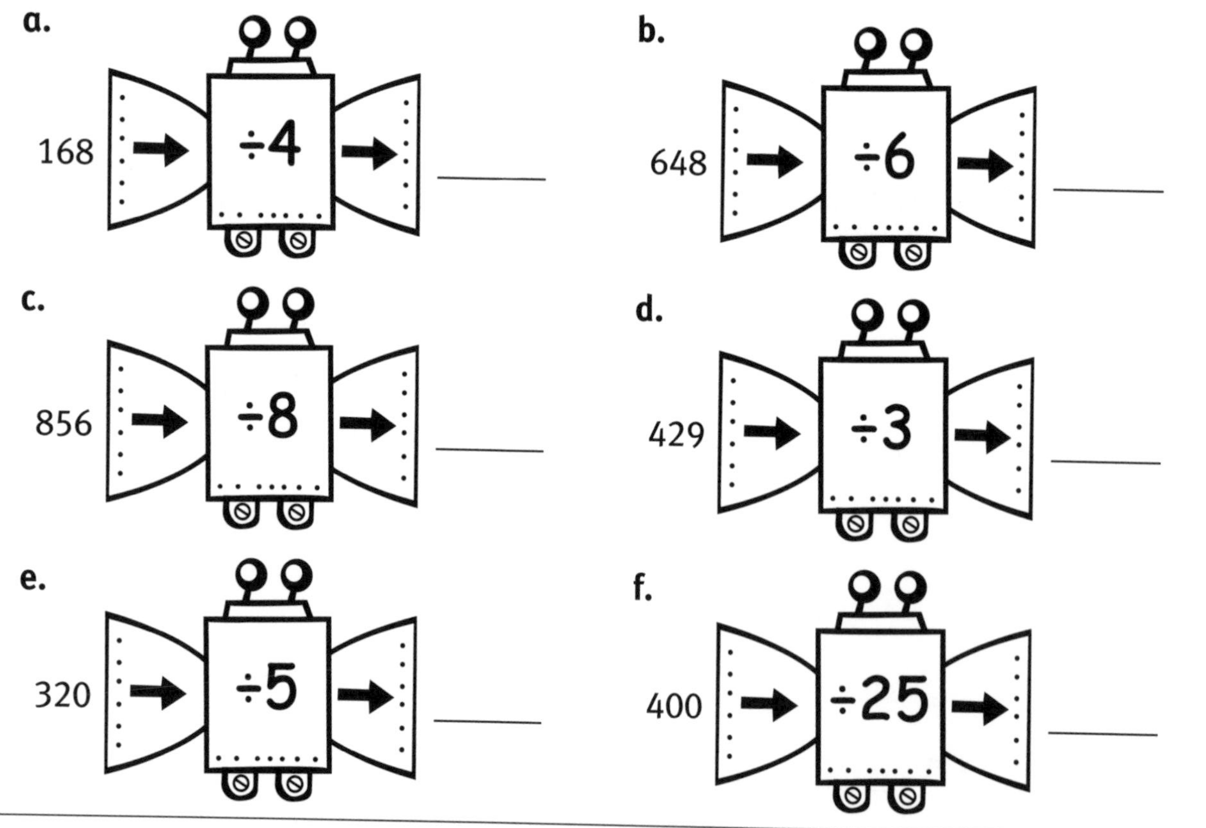

Name: ______________________

JUST FOR FUN 4

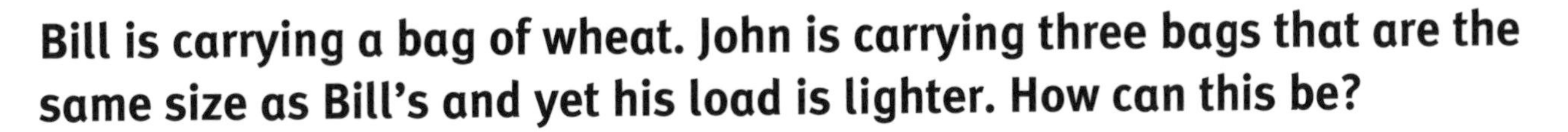

Bill is carrying a bag of wheat. John is carrying three bags that are the same size as Bill's and yet his load is lighter. How can this be?

Solve this riddle by figuring out the answers below. Write the letters above their matching answers at the bottom of the page. Some letters appear more than once.

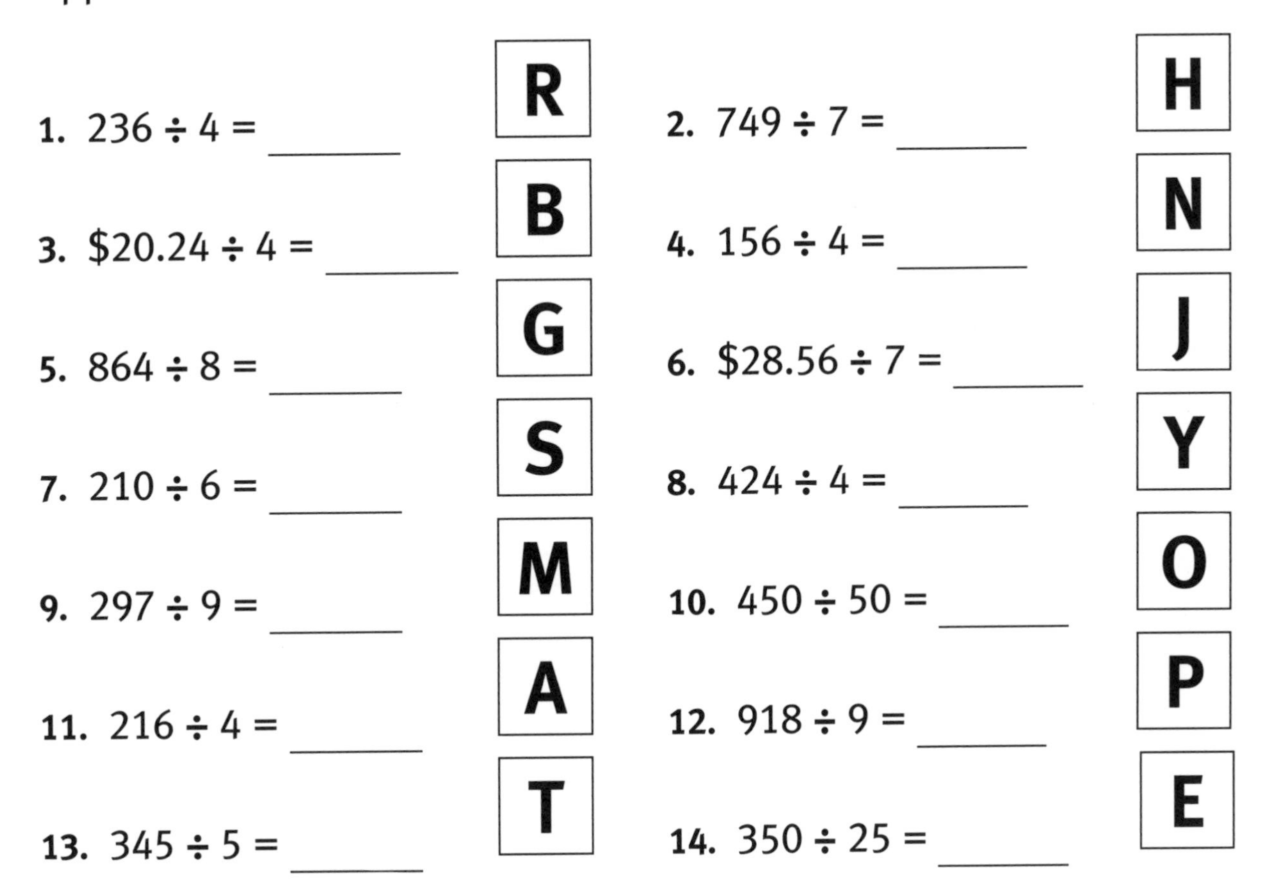

1. 236 ÷ 4 = ______ **R**
2. 749 ÷ 7 = ______ **H**
3. $20.24 ÷ 4 = ______ **B**
4. 156 ÷ 4 = ______ **N**
5. 864 ÷ 8 = ______ **G**
6. $28.56 ÷ 7 = ______ **J**
7. 210 ÷ 6 = ______ **S**
8. 424 ÷ 4 = ______ **Y**
9. 297 ÷ 9 = ______ **M**
10. 450 ÷ 50 = ______ **O**
11. 216 ÷ 4 = ______ **A**
12. 918 ÷ 9 = ______ **P**
13. 345 ÷ 5 = ______ **T**
14. 350 ÷ 25 = ______ **E**

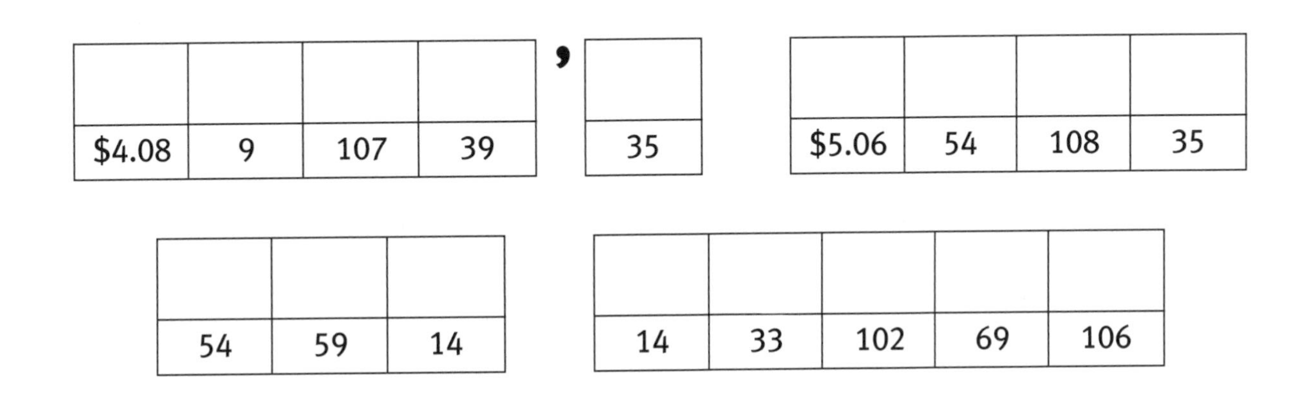

$4.08	9	107	39	, 35

$5.06	54	108	35

54	59	14

14	33	102	69	106

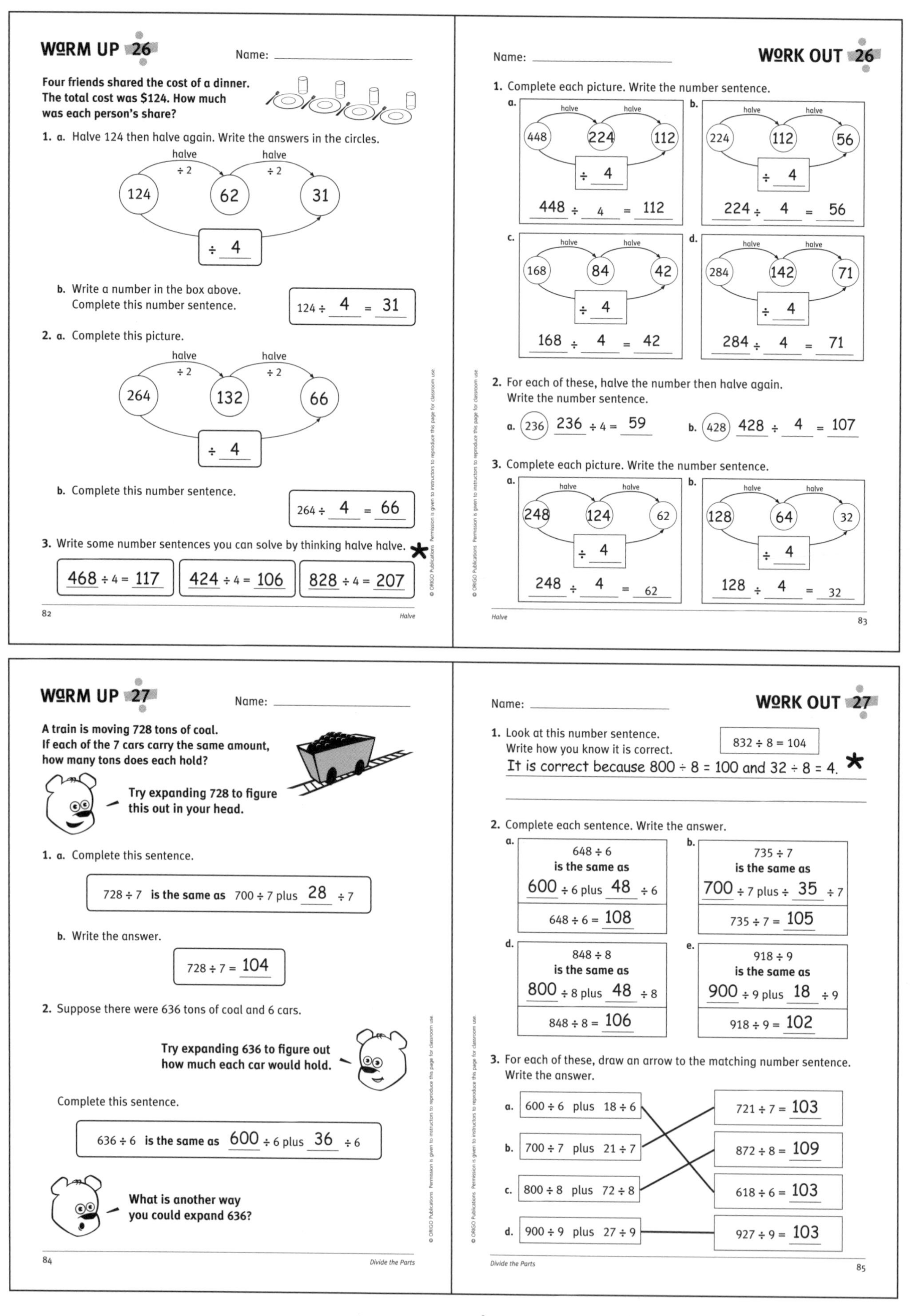

WORM UP 26

Name: ____________

Four friends shared the cost of a dinner. The total cost was $124. How much was each person's share?

1. a. Halve 124 then halve again. Write the answers in the circles.

halve ÷ 2 → halve ÷ 2: 124 → 62 → 31; ÷ 4

b. Write a number in the box above. Complete this number sentence.

124 ÷ 4 = 31

2. a. Complete this picture.

halve ÷ 2 → halve ÷ 2: 264 → 132 → 66; ÷ 4

b. Complete this number sentence.

264 ÷ 4 = 66

3. Write some number sentences you can solve by thinking halve halve. *

468 ÷ 4 = 117 | 424 ÷ 4 = 106 | 828 ÷ 4 = 207

82 — *Halve*

WORK OUT 26

Name: ____________

1. Complete each picture. Write the number sentence.

a. halve, halve: 448 → 224 → 112; ÷ 4. 448 ÷ 4 = 112

b. halve, halve: 224 → 112 → 56; ÷ 4. 224 ÷ 4 = 56

c. halve, halve: 168 → 84 → 42; ÷ 4. 168 ÷ 4 = 42

d. halve, halve: 284 → 142 → 71; ÷ 4. 284 ÷ 4 = 71

2. For each of these, halve the number then halve again. Write the number sentence.

a. (236) 236 ÷ 4 = 59

b. (428) 428 ÷ 4 = 107

3. Complete each picture. Write the number sentence.

a. halve, halve: 248 → 124 → 62; ÷ 4. 248 ÷ 4 = 62

b. halve, halve: 128 → 64 → 32; ÷ 4. 128 ÷ 4 = 32

Halve — 83

© ORIGO Publications Permission is given to instructors to reproduce this page for classroom use

WORM UP 27

Name: ____________

A train is moving 728 tons of coal. If each of the 7 cars carry the same amount, how many tons does each hold?

Try expanding 728 to figure this out in your head.

1. a. Complete this sentence.

728 ÷ 7 **is the same as** 700 ÷ 7 plus 28 ÷ 7

b. Write the answer.

728 ÷ 7 = 104

2. Suppose there were 636 tons of coal and 6 cars.

Try expanding 636 to figure out how much each car would hold.

Complete this sentence.

636 ÷ 6 **is the same as** 600 ÷ 6 plus 36 ÷ 6

What is another way you could expand 636?

84 — *Divide the Parts*

WORK OUT 27

Name: ____________

1. Look at this number sentence. Write how you know it is correct.

832 ÷ 8 = 104

It is correct because 800 ÷ 8 = 100 and 32 ÷ 8 = 4. *

2. Complete each sentence. Write the answer.

a. 648 ÷ 6 **is the same as** 600 ÷ 6 plus 48 ÷ 6. 648 ÷ 6 = 108

b. 735 ÷ 7 **is the same as** 700 ÷ 7 plus ÷ 35 ÷ 7. 735 ÷ 7 = 105

d. 848 ÷ 8 **is the same as** 800 ÷ 8 plus 48 ÷ 8. 848 ÷ 8 = 106

e. 918 ÷ 9 **is the same as** 900 ÷ 9 plus 18 ÷ 9. 918 ÷ 9 = 102

3. For each of these, draw an arrow to the matching number sentence. Write the answer.

a. 600 ÷ 6 plus 18 ÷ 6 → 618 ÷ 6 = 103

b. 700 ÷ 7 plus 21 ÷ 7 → 721 ÷ 7 = 103

c. 800 ÷ 8 plus 72 ÷ 8 → 872 ÷ 8 = 109

d. 900 ÷ 9 plus 27 ÷ 9 → 927 ÷ 9 = 103

Divide the Parts — 85

© ORIGO Publications Permission is given to instructors to reproduce this page for classroom use

* Answers will vary. This is one example.

WORM UP 28

Name: ______________

A 5-day pass costs $35.50.
What is the cost of one day's travel?

5-DAY Travel Pass

Try dividing the dollars and cents separately.

1. a. Complete this sentence.

$35.50 ÷ 5 is the same as $ 35 ÷ 5 plus 50 ¢ ÷ 5

b. Write the answer. $7.10

2. Suppose the pass was $25.15.

Figure out the cost of one day's travel by dividing the dollars then the cents.

a. Complete this sentence.

$25.15 ÷ 5 is the same as $ 25 ÷ 5 plus 15 ¢ ÷ 5

b. Write the answer. $5.03

How can you mentally check your answer?

WORK OUT 28

Name: ______________

1. Lenny figured that $24.32 ÷ 4 is $6.80.
What did he do wrong?
He divided 32¢ by 4 and wrote 80¢ instead of 8¢. *

2. For each of these, complete the sentence then write the answer.

a. $36.48 ÷ 4 is the same as $ 36 ÷ 4 plus 48 ¢ ÷ 4 = $9.12

b. $32.40 ÷ 8 is the same as $ 32 ÷ 8 plus 40 ¢ ÷ 8 = $4.05

c. $28.56 ÷ 7 is the same as $ 28 ÷ 7 plus 56 ¢ ÷ 7 = $4.08

d. $20.32 ÷ 4 is the same as $ 20 ÷ 4 plus 32 ¢ ÷ 4 = $5.08

3. Calculate the answers to these. Write number sentences.

a. A magazine subscription costs $27.69 for 3 issues. How much does 1 issue cost?
$27.69 ÷ 3 = $9.23

b. A special pack of 4 CDs costs $84.48. How much does one CD cost?
$84.48 ÷ 4 = $21.12

c. Six people shared a pizza dinner costing $72.60. How much did each person pay?
$72.60 ÷ 6 = $12.10

d. Four magazine issues cost $32.80. How much does each issue cost?
$32.80 ÷ 4 = $8.20

WORM UP 29

Name: ______________

A school bought 3 new baseball helmets. The total bill was $138. How much did each helmet cost?

Try breaking 138 into two smaller parts that are easier to divide by 3.

1. a. Complete this sentence to show your thinking.

138 ÷ 3 is the same as 120 ÷ 3 plus 18 ÷ 3

b. Write the answer. $46

2. Suppose the total bill was only $117.
Figure out the cost of each helmet by breaking 117 into smaller parts.

a. Complete this sentence.

117 ÷ 3 is the same as 90 ÷ 3 plus 27 ÷ 3

b. Write the answer. $39

What are some other ways you could calculate 138 ÷ 3 and 117 ÷ 3 in your head?

WORK OUT 29

Name: ______________

1. Suppose you had to divide each of these by 4.
Break each number into two parts that are easier to divide by 4.

a. 136 → 120 + 16
b. 156 * → 120 + 36
c. 176 * → 160 + 16
d. 260 * → 240 + 20

2. Break each of these into parts that you can easily divide by 6.

a. 156 * → 120 + 36
b. 492 * → 480 + 12
c. 450 * → 420 + 30
d. 216 * → 180 + 36

3. Figure out the answers to these by breaking the numbers into two smaller parts. Write the answers.

a. 228 ÷ 3 is the same as 210 ÷ 3 plus 18 ÷ 3 = 76

b. 344 ÷ 8 is the same as 320 ÷ 8 plus 24 ÷ 8 = 43

c. 297 ÷ 9 is the same as 270 ÷ 9 plus 27 ÷ 9 = 33

d. 161 ÷ 7 is the same as 140 ÷ 7 plus 21 ÷ 7 = 23

4. Use the same strategy to help you figure out each of these.
Write the answer.

a. 198 ÷ 9 = 22 b. 224 ÷ 7 = 32 c. 112 ÷ 8 = 14

d. 115 ÷ 5 = 23 e. 129 ÷ 3 = 43 f. 272 ÷ 8 = 34

* Answers will vary. This is one example.

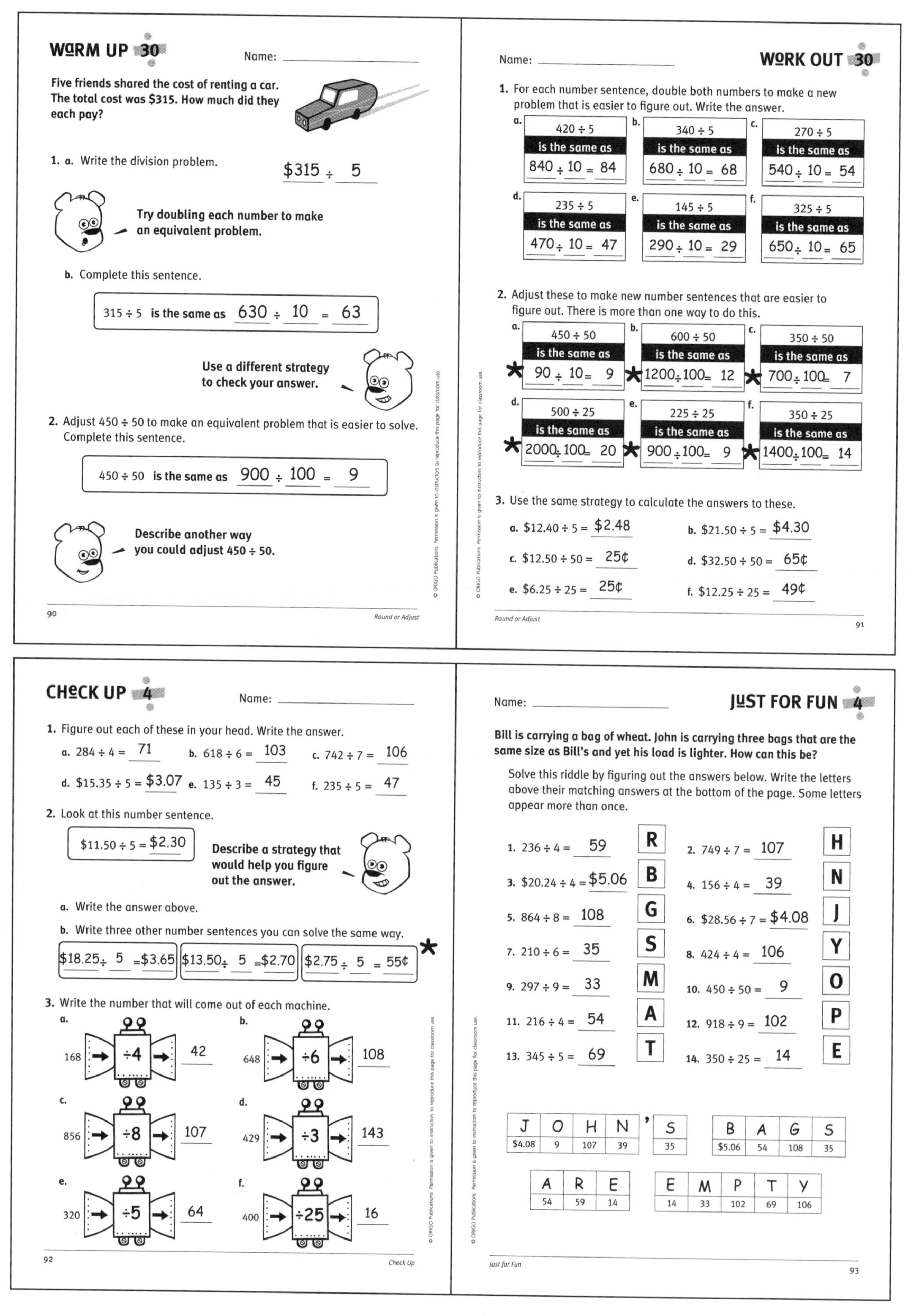

WARM UP 30

Name: ____________

Five friends shared the cost of renting a car. The total cost was $315. How much did they each pay?

1. a. Write the division problem. $315 ÷ 5

Try doubling each number to make an equivalent problem.

b. Complete this sentence.

315 ÷ 5 is the same as 630 ÷ 10 = 63

Use a different strategy to check your answer.

2. Adjust 450 ÷ 50 to make an equivalent problem that is easier to solve. Complete this sentence.

450 ÷ 50 is the same as 900 ÷ 100 = 9

Describe another way you could adjust 450 ÷ 50.

© ORIGO Publications Permission is given to instructors to reproduce this page for classroom use.

90 Round or Adjust

WORK OUT 30

Name: ____________

1. For each number sentence, double both numbers to make a new problem that is easier to figure out. Write the answer.

a. 420 ÷ 5 is the same as 840 ÷ 10 = 84
b. 340 ÷ 5 is the same as 680 ÷ 10 = 68
c. 270 ÷ 5 is the same as 540 ÷ 10 = 54
d. 235 ÷ 5 is the same as 470 ÷ 10 = 47
e. 145 ÷ 5 is the same as 290 ÷ 10 = 29
f. 325 ÷ 5 is the same as 650 ÷ 10 = 65

2. Adjust these to make new number sentences that are easier to figure out. There is more than one way to do this.

a. 450 ÷ 50 is the same as * 90 ÷ 10 = 9
b. 600 ÷ 50 is the same as * 1200 ÷ 100 = 12
c. 350 ÷ 50 is the same as * 700 ÷ 100 = 7
d. 500 ÷ 25 is the same as * 2000 ÷ 100 = 20
e. 225 ÷ 25 is the same as * 900 ÷ 100 = 9
f. 350 ÷ 25 is the same as * 1400 ÷ 100 = 14

3. Use the same strategy to calculate the answers to these.

a. $12.40 ÷ 5 = $2.48
b. $21.50 ÷ 5 = $4.30
c. $12.50 ÷ 50 = 25¢
d. $32.50 ÷ 50 = 65¢
e. $6.25 ÷ 25 = 25¢
f. $12.25 ÷ 25 = 49¢

© ORIGO Publications Permission is given to instructors to reproduce this page for classroom use.

Round or Adjust 91

CHECK UP 4

Name: ____________

1. Figure out each of these in your head. Write the answer.

a. 284 ÷ 4 = 71
b. 618 ÷ 6 = 103
c. 742 ÷ 7 = 106
d. $15.35 ÷ 5 = $3.07
e. 135 ÷ 3 = 45
f. 235 ÷ 5 = 47

2. Look at this number sentence.

$11.50 ÷ 5 = $2.30

Describe a strategy that would help you figure out the answer.

a. Write the answer above.

b. Write three other number sentences you can solve the same way. *

$18.25 ÷ 5 = $3.65 | $13.50 ÷ 5 = $2.70 | $2.75 ÷ 5 = 55¢

3. Write the number that will come out of each machine.

a. 168 → ÷4 → 42
b. 648 → ÷6 → 108
c. 856 → ÷8 → 107
d. 429 → ÷3 → 143
e. 320 → ÷5 → 64
f. 400 → ÷25 → 16

© ORIGO Publications Permission is given to instructors to reproduce this page for classroom use.

92 Check Up

JUST FOR FUN 4

Name: ____________

Bill is carrying a bag of wheat. John is carrying three bags that are the same size as Bill's and yet his load is lighter. How can this be?

Solve this riddle by figuring out the answers below. Write the letters above their matching answers at the bottom of the page. Some letters appear more than once.

1. 236 ÷ 4 = 59 **R**
2. 749 ÷ 7 = 107 **H**
3. $20.24 ÷ 4 = $5.06 **B**
4. 156 ÷ 4 = 39 **N**
5. 864 ÷ 8 = 108 **G**
6. $28.56 ÷ 7 = $4.08 **J**
7. 210 ÷ 6 = 35 **S**
8. 424 ÷ 4 = 106 **Y**
9. 297 ÷ 9 = 33 **M**
10. 450 ÷ 50 = 9 **O**
11. 216 ÷ 4 = 54 **A**
12. 918 ÷ 9 = 102 **P**
13. 345 ÷ 5 = 69 **T**
14. 350 ÷ 25 = 14 **E**

J	O	H	N	'	S
$4.08	9	107	39		35

B	A	G	S
$5.06	54	108	35

A	R	E
54	59	14

E	M	P	T	Y
14	33	102	69	106

© ORIGO Publications Permission is given to instructors to reproduce this page for classroom use.

Just for Fun 93

* Answers will vary. This is one example.